Numerical Recipes Example Book

(FORTRAN)

Numerical Recipes Example Book

(FORTRAN)

William T. Vetterling

Polaroid Corporation

Saul A. Teukolsky

Department of Physics, Cornell University

William H. Press

Harvard-Smithsonian Center for Astrophysics

Brian P. Flannery

EXXON Research and Engineering Company

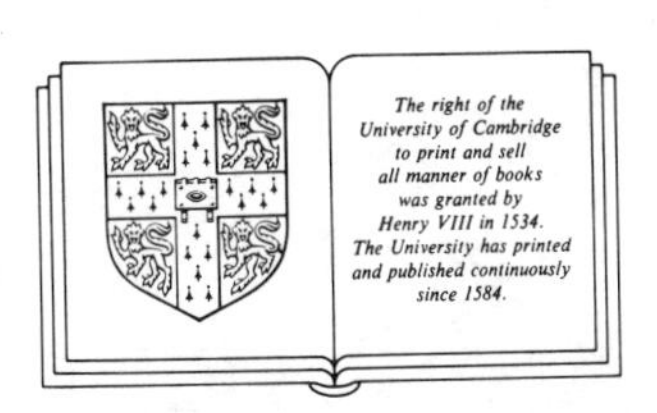

Cambridge University Press

Cambridge

London *New York*

New Rochelle *Melbourne* *Sydney*

Published by the Press Syndicate of the University of Cambridge
The Pitt Building, Trumpington Street, Cambridge CB2 1RP
32 East 57th Street, New York, NY 10022 U.S.A.
10 Stamford Road, Oakleigh, Melbourne 3166, Australia

First published 1985

Printed in the United States of America
Typeset in TEX

The computer programs in this book, and the subroutines in *Numerical Recipes: The Art of Scientific Computing*, are available in several machine-readable formats, in both the FORTRAN and Pascal programming languages. To purchase diskettes in IBM-compatible personal computer formats, use the order form at the back of this book, or write to Cambridge University Press, 510 North Avenue, New Rochelle, New York, 10801 or to Numerical Recipes Software, P.O. Box 243, Cambridge, Massachusetts 02238. For information on other formats, and on related software products, write to Numerical Recipes Software at the above address.

ISBN 0 521 31330 9

CONTENTS

Preface

This *Numerical Recipes Example Book* is designed to accompany the text and reference book *Numerical Recipes: The Art of Scientific Computing* by William H. Press, Brian P. Flannery, Saul A. Teukolsky, and William T. Vetterling (Cambridge University Press, 1985). In that volume, the algorithms and methods of scientific computation are developed in considerable detail, starting with basic mathematical analysis and working through to actual implementation in the form of FORTRAN subroutines and (in parallel) Pascal procedures. The routines in *Numerical Recipes: The Art of Scientific Computing*, numbering nearly 200, are meant to be incorporated into user applications; they are subroutines (or functions), not stand-alone programs.

It often happens, when you want to incorporate somebody else's subroutine into your own application program, that you first want to see the subroutine demonstrated on a simple example. Prose descriptions of how to use a subroutine (even those in *Numerical Recipes*) can occasionally be inexact. There is no substitute for an actual, FORTRAN demonstration program that shows exactly how data are fed to a subroutine, how the subroutine is called, and how its results are unloaded and interpreted.

Another not unusual case occurs when you have, for one seemingly good purpose or another, modified the source code in a "foreign" subroutine. In such circumstances, you might well want to test the modified subroutine on an example known previously to have worked correctly, *before* letting it loose on your own data. There is the related case where subroutine source code may have become corrupted, e.g., lost some lines or characters in transmission from one machine to another, and a simple revalidation test is desirable.

These are the needs addressed by this *Numerical Recipes Example Book.* Divided into chapters identically with *Numerical Recipes: The Art of Scientific Computing*, this book contains FORTRAN source programs that exercise and demonstrate all of the *Numerical Recipes* subroutines and functions. Each program contains comments, and is prefaced by a short description of what it does, and of which *Numerical Recipes* routines it exercises. In cases where the demonstration programs require input data, that data is also printed in this book. In some cases, where the demonstration programs are not "self-validating," sample output is also shown.

Necessarily, in the interests of clarity, the *Numerical Recipes* subroutines and functions are demonstrated in simple ways. A consequence is that the demonstration programs in this book do not usually test all possible regimes of input data, or even all lines of subroutine source code. The demonstration programs in this book were by no means the only validating tests that the *Numerical Recipes* subroutines and functions were required to pass during their development. The programs in this book *were* used during the later stages of the production of *Numerical Recipes: The Art of Scientific Computing* to maintain integrity of the source code, and in this role were found to be invaluable.

Disclaimer of Warranty

Chapter 1: Preliminaries

The routines in Chapter 1 of Numerical Recipes are introductory and less general in purpose than those in the remainder of the book. This chapter's routines serve primarily to expose the book's notational conventions, illustrate control structures, and perhaps to amuse. You may even find them useful. We hope that you will use BADLUK *for no serious purpose.*

⋆ ⋆ ⋆ ⋆

Subroutine FLMOON calculates the phases of the moon, or more exactly, the Julian day and fraction thereof on which a given phase will occur or has occurred. The program D1R1 asks the present date and compiles a list of upcoming phases. We have compared the predictions to lunar tables, with happy results. Shown are the results of a test run, which you may replicate as a check. In this program, notice that we have set TZONE (the time zone) to 5.0 to signify the five hour separation of the Eastern Standard time zone from Greenwich, England. Our parochial viewpoint requires you to use negative values of TZONE if you are east of Greenwich. The Julian day results are converted to calendar dates through the use of CALDAT, which appears later in the chapter. The fractional Julian day and time zone combine to form a correction that can possibly change the calendar date by one day.

Date	Time(EST)	Phase
1 9 1982	3 PM	full moon
1 16 1982	7 PM	last quarter
1 24 1982	11 PM	new moon
2 1 1982	10 AM	first quarter
2 8 1982	2 AM	full moon
2 15 1982	3 PM	last quarter
2 23 1982	4 PM	new moon
3 2 1982	6 PM	first quarter
3 9 1982	3 PM	full moon
3 17 1982	12 AM	last quarter
3 25 1982	5 AM	new moon
4 1 1982	0 AM	first quarter
4 8 1982	5 AM	full moon
4 16 1982	7 AM	last quarter
4 23 1982	4 PM	new moon
4 30 1982	7 AM	first quarter
5 7 1982	8 PM	full moon
5 16 1982	0 AM	last quarter
5 23 1982	0 AM	new moon
5 29 1982	3 PM	first quarter

```
      PROGRAM D1R1
C     Driver for routine FLMOON
      PARAMETER(TZONE=5.0)
      CHARACTER PHASE(4)*15,TIMSTR(2)*3
      DATA PHASE/'new moon','first quarter',
     *           'full moon','last quarter'/
      DATA TIMSTR/' AM',' PM'/
      WRITE(*,*) 'Date of the next few phases of the moon'
      WRITE(*,*) 'Enter today''s date (e.g. 1,31,1982)'
      TIMZON=-TZONE/24.0
      READ(*,*) IM,ID,IY
C     Approximate number of full moons since January 1900
      N=12.37*(IY-1900+(IM-0.5)/12.0)
      NPH=2
      J1=JULDAY(IM,ID,IY)
      CALL FLMOON(N,NPH,J2,FRAC)
      N=N+(J1-J2)/28.0
      WRITE(*,'(/1X,T6,A,T19,A,T32,A)') 'Date','Time(EST)','Phase'
      DO 11 I=1,20
          CALL FLMOON(N,NPH,J2,FRAC)
          IFRAC=NINT(24.*(FRAC+TIMZON))
          IF (IFRAC.LT.0) THEN
              J2=J2-1
              IFRAC=IFRAC+24
          ENDIF
          IF (IFRAC.GE.12) THEN
              J2=J2+1
              IFRAC=IFRAC-12
          ELSE
              IFRAC=IFRAC+12
          ENDIF
          IF (IFRAC.GT.12) THEN
              IFRAC=IFRAC-12
              ISTR=2
          ELSE
              ISTR=1
          ENDIF
          CALL CALDAT(J2,IM,ID,IY)
          WRITE(*,'(1X,2I3,I5,T20,I2,A,5X,A)') IM,ID,IY,
     *          IFRAC,TIMSTR(ISTR),PHASE(NPH+1)
          IF (NPH.EQ.3) THEN
              NPH=0
              N=N+1
          ELSE
              NPH=NPH+1
          ENDIF
11    CONTINUE
      END
```

Program JULDAY, our exemplar of the IF control structure, converts calendar dates to Julian dates. Not many people know the Julian date of their birthday or any other convenient reference point, for that matter. To remedy this, we offer a list of checkpoints, which appears at the end of this chapter as the file DATES.DAT. The program D1R2 lists the Julian date for each historic event for comparison. Then it allows you to make your own choices for entertainment.

```
      PROGRAM D1R2
C     Driver for JULDAY
      CHARACTER TXT*40,NAME(12)*15
      DATA NAME/'January','February','March','April','May','June',
     *        'July','August','September','October','November',
     *        'December'/
      OPEN(5,FILE='DATES.DAT',STATUS='OLD')
      READ(5,'(A)') TXT
      READ(5,*) N
      WRITE(*,'(/1X,A,T12,A,T17,A,T23,A,T37,A/)') 'Month','Day','Year',
     *        'Julian Day','Event'
      DO 11 I=1,N
          READ(5,'(I2,I3,I5,A)') IM,ID,IY,TXT
          WRITE(*,'(1X,A10,I3,I6,3X,I7,5X,A)') NAME(IM),ID,IY,
     *            JULDAY(IM,ID,IY),TXT
11    CONTINUE
      CLOSE(5)
      WRITE(*,'(/1X,A/)') 'Month,Day,Year (e.g. 1,13,1905)'
      DO 12 I=1,20
          WRITE(*,*) 'MM,DD,YYYY'
          READ(*,*) IM,ID,IY
          IF(IM.LT.0) STOP
          WRITE(*,'(1X,A12,I8/)') 'Julian Day: ',JULDAY(IM,ID,IY)
12    CONTINUE
      END
```

The next program in *Numerical Recipes* is BADLUK, an infamous code that combines the best and worst instincts of man. We include no demonstration program for BADLUK, not just because we fear it, but also because it is self-contained, with sample results appearing in the text.

Chapter 1 closes with routine CALDAT, which illustrates no new points, but complements JULDAY by doing conversions from Julian day number to the month, day, and year on which the given Julian day began. This offers an opportunity, grasped by the demonstration program D1R4, to push dates through both JULDAY and CALDAT in succession, to see if they survive intact. This, of course, tests only your authors' ability to make mistakes backwards as well as forwards, but we hope you will share our optimism that correct results here speak well for both routines. (We have checked them a bit more carefully in other ways.)

```
      PROGRAM D1R4
C     Driver for routine CALDAT
      CHARACTER NAME(12)*10
C     Check whether CALDAT properly undoes the operation of JULDAY
      DATA NAME/'January','February','March','April','May',
     *        'June','July','August','September','October',
     *        'November','December'/
      OPEN(5,FILE='DATES.DAT',STATUS='OLD')
      READ(5,*)
      READ(5,*) N
      WRITE(*,'(/1X,A,T40,A)') 'Original Date:','Reconstructed Date:'
      WRITE(*,'(1X,A,T12,A,T17,A,T25,A,T40,A,T50,A,T55,A/)')
     *        'Month','Day','Year','Julian Day','Month','Day','Year'
      DO 11 I=1,N
          READ(5,'(I2,I3,I5)') IM,ID,IY
          IYCOPY=IY
```

```
        J=JULDAY(IM,ID,IYCOPY)
        CALL CALDAT(J,IMM,IDD,IYY)
        WRITE(*,'(1X,A,I3,I6,4X,I9,6X,A,I3,I6)') NAME(IM),ID,
     *          IY,J,NAME(IMM),IDD,IYY
11    CONTINUE
      END
```

Appendix

File DATES.DAT:

```
List of dates for testing routines in Chapter 1
16 entries
12 31   -1 End of millennium
01 01    1 One day later
10 14 1582 Day before Gregorian calendar
10 15 1582 Gregorian calendar adopted
01 17 1706 Benjamin Franklin born
04 14 1865 Abraham Lincoln shot
04 18 1906 San Francisco earthquake
05 07 1915 Sinking of the Lusitania
07 20 1923 Pancho Villa assassinated
05 23 1934 Bonnie and Clyde eliminated
07 22 1934 John Dillinger shot
04 03 1936 Bruno Hauptman electrocuted
05 06 1937 Hindenburg disaster
07 26 1956 Sinking of the Andrea Doria
06 05 1976 Teton dam collapse
05 23 1968 Julian Day 2440000
```

Chapter 2: Linear Algebraic Equations

*Numerical Recipes Chapter 2 begins the "true grit" of numerical analysis by considering the solution of linear algebraic equations. This is done first by Gauss-Jordan elimination (*GAUSSJ*), and then by LU decomposition with forward and back substitution (*LUDCMP *and* LUBKSB*). For singular or nearly singular matrices the best choice is singular value decomposition with back substitution (*SVDCMP *and* SVBKSB*). Several linear systems of special form, represented by tridiagonal, Vandermonde, and Toeplitz matrices, may be treated with subroutines* TRIDAG, VANDER, *and* TOEPLZ *respectively. Linear systems with relatively few non-zero coefficients, so-called "sparse" matrices, are handled by routine* SPARSE.

⋆ ⋆ ⋆ ⋆

GAUSSJ performs Gauss-Jordan elimination with full pivoting to find the solution of a set of linear equations for a collection of right-hand side vectors. The demonstration routine D2R1 checks its operation with reference to a group of test input matrices printed at the end of this chapter as file MATRX1.DAT. Each matrix is subjected to inversion by GAUSSJ, and then multiplication by its own inverse to see that a unit matrix is produced. Then the solution vectors are each checked through multiplication by the original matrix and comparison with the right-hand side vectors that produced them.

```
      PROGRAM D2R1
C     Driver program for subroutine GAUSSJ
C     Reads Matrices from file MATRIX.DAT and feeds them to GAUSSJ
      PARAMETER(NP=20)
      DIMENSION A(NP,NP),B(NP,NP),AI(NP,NP),X(NP,NP)
      DIMENSION U(NP,NP),T(NP,NP)
      CHARACTER DUMMY*3
      OPEN(5,FILE='MATRX1.DAT',STATUS='OLD')
10    READ(5,'(A)') DUMMY
      IF (DUMMY.EQ.'END') GOTO 99
      READ(5,*)
      READ(5,*) N,M
      READ(5,*)
      READ(5,*) ((A(K,L), L=1,N), K=1,N)
      READ(5,*)
      READ(5,*) ((B(K,L), K=1,N), L=1,M)
C     Save Matrices for later testing of results
      DO 13 L=1,N
        DO 11 K=1,N
          AI(K,L)=A(K,L)
11      CONTINUE
```

```
          DO 12 K=1,M
            X(L,K)=B(L,K)
12        CONTINUE
13      CONTINUE
C       Invert Matrix
        CALL GAUSSJ(AI,N,NP,X,M,NP)
        WRITE(*,*) 'Inverse of Matrix A : '
        DO 14 K=1,N
          WRITE(*,'(1H,(6F12.6))') (AI(K,L), L=1,N)
14      CONTINUE
C       Test Results
C       Check Inverse
        WRITE(*,*) 'A times A-inverse (compare with unit matrix)'
        DO 17 K=1,N
          DO 16 L=1,N
            U(K,L)=0.0
            DO 15 J=1,N
              U(K,L)=U(K,L)+A(K,J)*AI(J,L)
15          CONTINUE
16        CONTINUE
          WRITE(*,'(1H,(6F12.6))') (U(K,L), L=1,N)
17      CONTINUE
C       Check Vector Solutions
        WRITE(*,*) 'Check the following vectors for equality:'
        WRITE(*,'(T12,A8,T23,A12)') 'Original','Matrix*Sol''n'
        DO 21 L=1,M
            WRITE(*,'(1X,A,I2,A)') 'Vector ',L,':'
          DO 19 K=1,N
            T(K,L)=0.0
            DO 18 J=1,N
              T(K,L)=T(K,L)+A(K,J)*X(J,L)
18          CONTINUE
          WRITE(*,'(8X,2F12.6)') B(K,L),T(K,L)
19        CONTINUE
21      CONTINUE
        WRITE(*,*) '***********************************'
        WRITE(*,*) 'Press RETURN for next problem:'
        READ(*,*)
        GOTO 10
99      CLOSE(5)
        END
```

The demonstration program for routine LUDCMP relies on the same package of test matrices, but just performs an LU decomposition of each. The performance is checked by multiplying the lower and upper matrices of the decomposition and comparing with the original matrix. The array INDX keeps track of the scrambling done by LUDCMP to effect partial pivoting. We had to do the unscrambling here, but you will normally not be called upon to do so, since LUDCMP is used with the descrambler-containing routine LUBKSB.

```
        PROGRAM D2R2
C       Driver for routine LUDCMP
        PARAMETER(NP=20)
        DIMENSION A(NP,NP),XL(NP,NP),XU(NP,NP)
        DIMENSION INDX(NP),JNDX(NP),X(NP,NP)
        CHARACTER TXT*3
```

```
      OPEN (5,FILE='MATRX1.DAT',STATUS='OLD')
      READ(5,*)
10    READ(5,*)
      READ(5,*) N,M
      READ(5,*)
      READ(5,*) ((A(K,L), L=1,N), K=1,N)
      READ(5,*)
      READ(5,*) ((X(K,L), K=1,N), L=1,M)
C     Print out A-matrix for comparison with product of lower
C     and upper decomposition matrices.
      WRITE(*,*) 'Original matrix:'
      DO 11 K=1,N
        WRITE(*,'(1X,6F12.6)') (A(K,L), L=1,N)
11    CONTINUE
C     Perform the decomposition
      CALL LUDCMP(A,N,NP,INDX,D)
C     Compose separately the lower and upper matrices
      DO 13 K=1,N
        DO 12 L=1,N
          IF (L.GT.K) THEN
            XU(K,L)=A(K,L)
            XL(K,L)=0.0
          ELSE IF (L.LT.K) THEN
            XU(K,L)=0.0
            XL(K,L)=A(K,L)
          ELSE
            XU(K,L)=A(K,L)
            XL(K,L)=1.0
          ENDIF
12      CONTINUE
13    CONTINUE
C     Compute product of lower and upper matrices for
C     comparison with original matrix.
      DO 16 K=1,N
        JNDX(K)=K
        DO 15 L=1,N
          X(K,L)=0.0
          DO 14 J=1,N
            X(K,L)=X(K,L)+XL(K,J)*XU(J,L)
14        CONTINUE
15      CONTINUE
16    CONTINUE
      WRITE(*,*) 'Product of lower and upper matrices (unscrambled):'
      DO 17 K=1,N
        DUM=JNDX(INDX(K))
        JNDX(INDX(K))=JNDX(K)
        JNDX(K)=DUM
17    CONTINUE
      DO 19 K=1,N
        DO 18 J=1,N
          IF (JNDX(J).EQ.K) THEN
            WRITE(*,'(1X,6F12.6)') (X(J,L), L=1,N)
          ENDIF
18      CONTINUE
19    CONTINUE
      WRITE(*,*) 'Lower matrix of the decomposition:'
      DO 21 K=1,N
```

```
            WRITE(*,'(1X,6F12.6)') (XL(K,L), L=1,N)
21        CONTINUE
          WRITE(*,*) 'Upper matrix of the decomposition:'
          DO 22 K=1,N
            WRITE(*,'(1X,6F12.6)') (XU(K,L), L=1,N)
22        CONTINUE
          WRITE(*,*) '***********************************'
          WRITE(*,*) 'Press RETURN for next problem:'
          READ(*,*)
          READ(5,'(A3)') TXT
          IF (TXT.NE.'END') GOTO 10
          CLOSE(5)
          END
```

Our example driver for LUBKSB makes calls to both LUDCMP and LUBKSB in order to solve the linear equation problems posed in file MATRX1.DAT (see discussion of GAUSSJ). The original matrix of coefficients is applied to the solution vectors to check that the result matches the right-hand side vectors posed for each problem. We apologize for using routine LUDCMP in a test of LUBKSB, but LUDCMP has been tested independently, and anyway, LUBKSB is nothing without this partner program, so a test of the combination is more to the point.

```
          PROGRAM D2R3
C         Driver for routine LUBKSB
          PARAMETER (NP=20)
          DIMENSION A(NP,NP),B(NP,NP),INDX(NP)
          DIMENSION C(NP,NP),X(NP)
          CHARACTER TXT*3
          OPEN(5,FILE='MATRX1.DAT',STATUS='OLD')
          READ(5,*)
10        READ(5,*)
          READ(5,*) N,M
          READ(5,*)
          READ(5,*) ((A(K,L), L=1,N), K=1,N)
          READ(5,*)
          READ(5,*) ((B(K,L), K=1,N), L=1,M)
C         Save matrix A for later testing
          DO 12 L=1,N
            DO 11 K=1,N
              C(K,L)=A(K,L)
11          CONTINUE
12        CONTINUE
C         Do LU decomposition
          CALL LUDCMP(C,N,NP,INDX,P)
C         Solve equations for each right-hand vector
          DO 16 K=1,M
            DO 13 L=1,N
              X(L)=B(L,K)
13          CONTINUE
            CALL LUBKSB(C,N,NP,INDX,X)
C         Test results with original matrix
            WRITE(*,*) 'Right-hand side vector:'
            WRITE(*,'(1X,6F12.6)') (B(L,K), L=1,N)
            WRITE(*,*) 'Result of matrix applied to sol''n vector'
            DO 15 L=1,N
              B(L,K)=0.0
```

```
                DO 14 J=1,N
                    B(L,K)=B(L,K)+A(L,J)*X(J)
14              CONTINUE
15          CONTINUE
            WRITE(*,'(1X,6F12.6)') (B(L,K), L=1,N)
            WRITE(*,*) '***********************************'
16      CONTINUE
        WRITE(*,*) 'Press RETURN for next problem:'
        READ(*,*)
        READ(5,'(A3)') TXT
        IF (TXT.NE.'END') GOTO 10
        CLOSE(5)
        END
```

Subroutine TRIDAG solves linear equations with coefficients that form a tridiagonal matrix. We provide at the end of this chapter a second file of matrices MATRIX2.DAT for the demonstration driver. In all other respects, the demonstration program D2R4 operates in the same fashion as D2R3.

```
      PROGRAM D2R4
C     Driver for routine TRIDAG
      PARAMETER (NP=20)
      DIMENSION DIAG(NP),SUPERD(NP),SUBD(NP),RHS(NP),U(NP)
      CHARACTER TXT*3
      OPEN(UNIT=5,FILE='MATRX2.DAT',STATUS='OLD')
10    READ(5,'(A3)') TXT
      IF (TXT.EQ.'END') GOTO 99
      READ(5,*)
      READ(5,*) N
      READ(5,*)
      READ(5,*) (DIAG(K), K=1,N)
      READ(5,*)
      READ(5,*) (SUPERD(K), K=1,N-1)
      READ(5,*)
      READ(5,*) (SUBD(K), K=2,N)
      READ(5,*)
      READ(5,*) (RHS(K), K=1,N)
C     Carry out solution
      CALL TRIDAG(SUBD,DIAG,SUPERD,RHS,U,N)
      WRITE(*,*) 'The solution vector is:'
      WRITE(*,'(1X,6F12.6)') (U(K), K=1,N)
C     Test solution
      WRITE(*,*) '(matrix)*(sol''n vector) should be:'
      WRITE(*,'(1X,6F12.6)') (RHS(K), K=1,N)
      WRITE(*,*) 'Actual result is:'
      DO 11 K=1,N
        IF (K.EQ.1) THEN
          RHS(K)=DIAG(1)*U(1) + SUPERD(1)*U(2)
        ELSE IF (K.EQ.N) THEN
          RHS(K)=SUBD(N)*U(N-1) + DIAG(N)*U(N)
        ELSE
          RHS(K)=SUBD(K)*U(K-1) + DIAG(K)*U(K)
     *                  + SUPERD(K)*U(K+1)
        ENDIF
11    CONTINUE
      WRITE(*,'(1X,6F12.6)') (RHS(K), K=1,N)
      WRITE(*,*) '***********************************'
```

```
      WRITE(*,*) 'Press RETURN for next problem:'
      READ(*,*)
      GOTO 10
99    CLOSE(5)
      END
```

MPROVE is a short routine for improving the solution vector for a set of linear equations, providing that an LU decomposition has been performed on the matrix of coefficients. Our test of this function is to use LUDCMP and LUBKSB to solve a set of equations specified in the DATA statements at the beginning of the program. The solution vector is then corrupted by the addition of random values to each component. MPROVE works on the corrupted vector to recover the original.

```
      PROGRAM D2R5
C     Driver for routine MPROVE
      PARAMETER(N=5,NP=5)
      DIMENSION A(NP,NP),INDX(N),B(N),X(N),AA(NP,NP)
      DATA A/1.0,2.0,1.0,4.0,5.0,2.0,3.0,1.0,5.0,1.0,
     *        3.0,4.0,1.0,1.0,2.0,4.0,5.0,1.0,2.0,3.0,
     *        5.0,1.0,1.0,3.0,4.0/
      DATA B/1.0,1.0,1.0,1.0,1.0/
      DO 12 I=1,N
        X(I)=B(I)
        DO 11 J=1,N
          AA(I,J)=A(I,J)
11      CONTINUE
12    CONTINUE
      CALL LUDCMP(AA,N,NP,INDX,D)
      CALL LUBKSB(AA,N,NP,INDX,X)
      WRITE(*,'(/1X,A)') 'Solution vector for the equations:'
      WRITE(*,'(1X,5F12.6)') (X(I),I=1,N)
C     Now phoney up X and let MPROVE fit it
      IDUM=-13
      DO 13 I=1,N
        X(I)=X(I)*(1.0+0.2*RAN3(IDUM))
13    CONTINUE
      WRITE(*,'(/1X,A)') 'Solution vector with noise added:'
      WRITE(*,'(1X,5F12.6)') (X(I),I=1,N)
      CALL MPROVE(A,AA,N,NP,INDX,B,X)
      WRITE(*,'(/1X,A)') 'Solution vector recovered by MPROVE:'
      WRITE(*,'(1X,5F12.6)') (X(I),I=1,N)
      END
```

Vandermonde matrices of dimension $N \times N$ have elements that are entirely integer powers of N arbitrary numbers $x_1 \ldots x_N$. (See *Numerical Recipes* for details). In the demonstration program D2R6 we provide five such numbers to specify a 5×5 matrix, and five elements of a right-hand side vector Q. Routine VANDER is used to find the solution vector W. This vector is tested by applying the matrix to W and comparing the result to Q.

```
      PROGRAM D2R6
C     Driver for routine VANDER
      PARAMETER(N=5)
      DIMENSION X(N),Q(N),W(N),TERM(N)
      DATA X/1.0,1.5,2.0,2.5,3.0/
      DATA Q/1.0,1.5,2.0,2.5,3.0/
```

```
      CALL VANDER(X,W,Q,N)
      WRITE(*,*) 'Solution vector:'
      DO 11 I=1,N
        WRITE(*,'(5X,A2,I1,A4,E12.6)') 'W(',I,') = ',W(I)
11    CONTINUE
      WRITE(*,'(/1X,A)') 'Test of solution vector:'
      WRITE(*,'(1X,T6,A,T19,A)') 'mtrx*sol''n','original'
      SUM=0.0
      DO 12 I=1,N
        TERM(I)=W(I)
        SUM=SUM+W(I)
12    CONTINUE
      WRITE(*,'(1X,2F12.4)') SUM,Q(1)
      DO 14 I=2,N
        SUM=0.0
        DO 13 J=1,N
          TERM(J)=TERM(J)*X(J)
          SUM=SUM+TERM(J)
13      CONTINUE
        WRITE(*,'(1X,2F12.4)') SUM,Q(I)
14    CONTINUE
      END
```

A very similar test is applied to TOEPLZ, which operates on Toeplitz matrices. The $N \times N$ Toeplitz matrix is specified by $2N - 1$ numbers r_i, in this case taken to be simply a linear progression of values. A right-hand side y_i is chosen likewise. TOEPLZ finds the solution vector x_i, and checks it in the usual fashion.

```
      PROGRAM D2R7
C     Driver for routine TOEPLZ
      PARAMETER(N=5,N2=2*N)
      DIMENSION X(N),Y(N),R(N2)
      DO 11 I=1,N
        Y(I)=0.1*I
11    CONTINUE
      DO 12 I=1,2*N-1
        R(I)=0.1*I
12    CONTINUE
      CALL TOEPLZ(R,X,Y,N)
      WRITE(*,*) 'Solution vector:'
      DO 13 I=1,N
        WRITE(*,'(5X,A2,I1,A4,E13.6)') 'X(',I,') = ',X(I)
13    CONTINUE
      WRITE(*,'(/1X,A)') 'Test of solution:'
      WRITE(*,'(1X,T6,A,T19,A)') 'mtrx*soln','original'
      DO 15 I=1,N
        SUM=0.0
        DO 14 J=1,N
          SUM=SUM+R(N+I-J)*X(J)
14      CONTINUE
        WRITE(*,'(1X,2F12.4)') SUM,Y(I)
15    CONTINUE
      END
```

The pair SVDCMP, SVBKSB are tested in the same manner as LUDCMP, LUBKSB. That is, SVDCMP is checked independently to see that it yields proper decomposition of matrices. Then the pair of programs is tested as a unit to see that they provide correct solutions to some linear sets. (Note: Because of the order of programs in *Numerical Recipes*, the test of the pair in this case comes first). The matrices and solution vectors are given in the Appendix as file MATRX3.DAT.

Driver D2R8 brings in matrices A and right-hand side vectors B from MATRX3.DAT. Matrix A, itself, is saved for later use. It is copied into matrix U for processing by SVDCMP. The results of the processing are the three arrays U, W, V which form the singular value decomposition of A. The right-hand side vectors are fed one at a time to vector C, and the resulting solution vectors X are checked for accuracy through application of the saved matrix A.

```
      PROGRAM D2R8
C     Driver for routine SVBKSB, which calls routine SVDCMP
      PARAMETER(NP=20)
      DIMENSION A(NP,NP),B(NP,NP),U(NP,NP),W(NP)
      DIMENSION V(NP,NP),C(NP),X(NP)
      CHARACTER DUMMY*3
      OPEN(5,FILE='MATRX1.DAT',STATUS='OLD')
10    READ(5,'(A)') DUMMY
      IF (DUMMY.EQ.'END') GOTO 99
      READ(5,*)
      READ(5,*) N,M
      READ(5,*)
      READ(5,*) ((A(K,L), L=1,N), K=1,N)
      READ(5,*)
      READ(5,*) ((B(K,L), K=1,N), L=1,M)
C     Copy A into U
      DO 12 K=1,N
        DO 11 L=1,N
          U(K,L)=A(K,L)
11      CONTINUE
12    CONTINUE
C     Decompose matrix A
      CALL SVDCMP(U,N,N,NP,NP,W,V)
C     Find maximum singular value
      WMAX=0.0
      DO 13 K=1,N
        IF (W(K).GT.WMAX) WMAX=W(K)
13    CONTINUE
C     Define "small"
      WMIN=WMAX*(1.0E-6)
C     Zero the "small" singular values
      DO 14 K=1,N
        IF (W(K).LT.WMIN) W(K)=0.0
14    CONTINUE
C     Backsubstitute for each right-hand side vector
      DO 18 L=1,M
        WRITE(*,'(1X,A,I2)') 'Vector number ',L
        DO 15 K=1,N
          C(K)=B(K,L)
15      CONTINUE
        CALL SVBKSB(U,W,V,N,N,NP,NP,C,X)
        WRITE(*,*) '    Solution vector is:'
        WRITE(*,'(1X,6F12.6)') (X(K), K=1,N)
        WRITE(*,*) '    Original right-hand side vector:'
```

```
            WRITE(*,'(1X,6F12.6)') (C(K), K=1,N)
            WRITE(*,*) '     Result of (matrix)*(sol''n vector):'
            DO 17 K=1,N
                C(K)=0.0
                DO 16 J=1,N
                    C(K)=C(K)+A(K,J)*X(J)
16              CONTINUE
17          CONTINUE
            WRITE(*,'(1X,6F12.6)') (C(K), K=1,N)
18      CONTINUE
        WRITE(*,*) '***********************************'
        WRITE(*,*) 'Press RETURN for next problem'
        READ(*,*)
        GOTO 10
99      CLOSE(5)
        END
```

Companion driver D2R9 takes the same matrices from MATRX3.DAT and passes copies U to SVDCMP for singular value decomposition into U, W, and V. Then U, W, and the transpose of V are multiplied together. The result is compared to a saved copy of A.

```
        PROGRAM D2R9
C       Driver for routine SVDCMP
        PARAMETER(NP=20)
        DIMENSION A(NP,NP),U(NP,NP),W(NP),V(NP,NP)
        CHARACTER DUMMY*3
        OPEN(5,FILE='MATRX3.DAT',STATUS='OLD')
10      READ(5,'(A)') DUMMY
        IF (DUMMY.EQ.'END') GOTO 99
        READ(5,*)
        READ(5,*) M,N
        READ(5,*)
C       Copy original matrix into U
        DO 12 K=1,M
            READ(5,*) (A(K,L), L=1,N)
            DO 11 L=1,N
                U(K,L)=A(K,L)
11          CONTINUE
12      CONTINUE
        IF (N.GT.M) THEN
            DO 14 K=M+1,N
                DO 13 L=1,N
                    A(K,L)=0.0
                    U(K,L)=0.0
13              CONTINUE
14          CONTINUE
            M=N
        ENDIF
C       Perform decomposition
        CALL SVDCMP(U,M,N,NP,NP,W,V)
C       Print results
        WRITE(*,*) 'Decomposition Matrices:'
        WRITE(*,*) 'Matrix U'
        DO 15 K=1,M
            WRITE(*,'(1X,6F12.6))') (U(K,L),L=1,N)
15      CONTINUE
```

```
      WRITE(*,*) 'Diagonal of Matrix W'
      WRITE(*,'(1X,6F12.6))') (W(K),K=1,N)
      WRITE(*,*) 'Matrix V-Transpose'
      DO 16 K=1,N
         WRITE(*,'(1X,6F12.6))') (V(L,K),L=1,N)
16    CONTINUE
      WRITE(*,*) 'Check product against original matrix:'
      WRITE(*,*) 'Original Matrix:'
      DO 17 K=1,M
         WRITE(*,'(1X,6F12.6))') (A(K,L),L=1,N)
17    CONTINUE
      WRITE(*,*) 'Product U*W*(V-Transpose):'
      DO 21 K=1,M
         DO 19 L=1,N
            A(K,L)=0.0
            DO 18 J=1,N
               A(K,L)=A(K,L)+U(K,J)*W(J)*V(L,J)
18          CONTINUE
19       CONTINUE
         WRITE(*,'(1X,6F12.6))') (A(K,L),L=1,N)
21    CONTINUE
      WRITE(*,*) '***********************************'
      WRITE(*,*) 'Press RETURN for next problem'
      READ(*,*)
      GOTO 10
99    CLOSE(5)
      END
```

Routine SPARSE solves linear systems $\mathbf{A} \cdot \mathbf{x} = \mathbf{b}$ with a sparse matrix $\mathbf{A}$. Rather than specifying the entire matrix $\mathbf{A}$ (most elements of which are zero), the program calls two subroutines ASUB and ATSUB which are, for any input vector XIN, supposed to return the result XOUT of applying $\mathbf{A}$ and its transpose to XIN, respectively. In our sample program we define these two subroutines to implement the 20×20 matrix

$$\begin{pmatrix} 1.0 & 2.0 & 0.0 & 0.0 & \ldots \\ -2.0 & 1.0 & 2.0 & 0.0 & \ldots \\ 0.0 & -2.0 & 1.0 & 2.0 & \ldots \\ 0.0 & 0.0 & -2.0 & 1.0 & \ldots \\ \vdots & \vdots & \vdots & \vdots & \ddots \end{pmatrix}$$

As a right-hand side vector $\mathbf{b}$ we have taken $(3.0, 1.0, 1.0, \ldots, -1.0)$, and the solution is given as $\mathbf{x}$. Notice that the components of $\mathbf{x}$ are all initialized to zero. You will set them to some initial guess of the solution to your own problem, but this guess will usually suffice. The solution in D2R10 is given the usual checks.

```
      PROGRAM D2R10
C     Driver for SPARSE
      PARAMETER(N=20)
      COMMON M
      DIMENSION B(N),X(N),BCMP(N)
      EXTERNAL ASUB,ATSUB
      M=N
      DO 11 I=1,N
         X(I)=0.0
         B(I)=1.0
```

```
11      CONTINUE
        B(1)=3.0
        B(N)=-1.0
        CALL SPARSE(B,N,ASUB,ATSUB,X,RSQ)
        WRITE(*,'(/1X,A,E15.6)') 'Sum-squared residual:',RSQ
        WRITE(*,'(/1X,A)') 'Solution vector:'
        WRITE(*,'(1X,5F12.6)') (X(I),I=1,N)
        CALL ASUB(X,BCMP)
        WRITE(*,'(/1X,A)') 'press RETURN to continue...'
        READ(*,*)
        WRITE(*,'(1X,A/T8,A,T22,A)') 'Test of solution vector:','a*x','b'
        DO 12 I=1,N
            WRITE(*,'(1X,2F12.6)') BCMP(I),B(I)
12      CONTINUE
        END
        SUBROUTINE ASUB(XIN,XOUT)
        COMMON N
        DIMENSION XIN(N),XOUT(N)
        XOUT(1)=XIN(1)+2.0*XIN(2)
        XOUT(N)=-2.0*XIN(N-1)+XIN(N)
        DO 11 I=2,N-1
            XOUT(I)=-2.0*XIN(I-1)+XIN(I)+2.0*XIN(I+1)
11      CONTINUE
        RETURN
        END
        SUBROUTINE ATSUB(XIN,XOUT)
        COMMON N
        DIMENSION XIN(N),XOUT(N)
        XOUT(1)=XIN(1)-2.0*XIN(2)
        XOUT(N)=2.0*XIN(N-1)+XIN(N)
        DO 11 I=2,N-1
            XOUT(I)=2.0*XIN(I-1)+XIN(I)-2.0*XIN(I+1)
11      CONTINUE
        RETURN
        END
```

Appendix

File MATRX1.DAT:

```
MATRICES FOR INPUT TO TEST ROUTINES
Size of matrix (NxN), Number of solutions:
3,2
Matrix A:
1.0 0.0 0.0
0.0 2.0 0.0
0.0 0.0 3.0
Solution vectors:
1.0 0.0 0.0
1.0 1.0 1.0
NEXT PROBLEM
Size of matrix (NxN), Number of solutions:
3,2
Matrix A:
1.0 2.0 3.0
2.0 2.0 3.0
3.0 3.0 3.0
```

```
Solution vectors:
1.0 1.0 1.0
1.0 2.0 3.0
NEXT PROBLEM:
Size of matrix (NxN), Number of solutions:
5,2
Matrix A:
1.0 2.0 3.0 4.0 5.0
2.0 3.0 4.0 5.0 1.0
3.0 4.0 5.0 1.0 2.0
4.0 5.0 1.0 2.0 3.0
5.0 1.0 2.0 3.0 4.0
Solution vectors:
1.0 1.0 1.0 1.0 1.0
1.0 2.0 3.0 4.0 5.0
NEXT PROBLEM:
Size of matrix (NxN), Number of solutions:
5,2
Matrix A:
1.4 2.1 2.1 7.4 9.6
1.6 1.5 1.1 0.7 5.0
3.8 8.0 9.6 5.4 8.8
4.6 8.2 8.4 0.4 8.0
2.6 2.9 0.1 9.6 7.7
Solution vectors:
1.1 1.6 4.7 9.1 0.1
4.0 9.3 8.4 0.4 4.1
END
```

File MATRX2.DAT:

```
TRIDIAGONAL MATRICES FOR PROGRAM 'TRIDAG'
Dimension of matrix
3
Diagonal elements (N)
1.0 2.0 3.0
Super-diagonal elements (N-1)
2.0 3.0
Sub-diagonal elements (N-1)
2.0 3.0
Right-hand side vector (N)
1.0 2.0 3.0
NEXT PROBLEM:
Dimension of matrix
5
Diagonal elements (N)
1.0 1.0 1.0 1.0 1.0
Super-diagonal elements (N-1)
1.0 2.0 3.0 4.0
Sub-diagonal elements (N-1)
2.0 3.0 4.0 5.0
Right-hand side vector (N)
1.0 2.0 3.0 4.0 5.0
NEXT PROBLEM:
Dimension of matrix
5
Diagonal elements (N)
```

```
1.0 2.0 3.0 4.0 5.0
Super-diagonal elements (N-1)
2.0 3.0 4.0 5.0
Sub-diagonal elements (N-1)
2.0 3.0 4.0 5.0
Right-hand side vector (N)
1.0 1.0 1.0 1.0 1.0
NEXT PROBLEM:
Dimension of matrix
6
Diagonal elements (N)
9.7 9.5 5.2 3.5 5.1 6.0
Super-diagonal elements (N-1)
6.0 1.2 0.7 3.0 1.5
Sub-diagonal elements (N-1)
2.1 9.4 3.3 7.5 8.8
Right-hand side vector (N)
2.0 7.5 0.6 7.4 9.8 8.8
END
```

File MATRX3.DAT:

```
TEST MATRICES FOR SVDCMP:
Number of Rows, Columns
5,3
Matrix
1.0 2.0 3.0
2.0 3.0 4.0
3.0 4.0 5.0
4.0 5.0 6.0
5.0 6.0 7.0
NEXT PROBLEM:
Number of Rows, Columns
5,5
Matrix
1.0 2.0 3.0 4.0 5.0
2.0 2.0 3.0 4.0 5.0
3.0 3.0 3.0 4.0 5.0
4.0 4.0 4.0 4.0 5.0
5.0 5.0 5.0 5.0 5.0
NEXT PROBLEM:
Number of Rows, Columns
6,6
Matrix
3.0 5.3 5.6 3.5 6.8 5.7
0.4 8.2 6.7 1.9 2.2 5.3
7.8 8.3 7.7 3.3 1.9 4.8
5.5 8.8 3.0 1.0 5.1 6.4
5.1 5.1 3.6 5.8 5.7 4.9
3.5 2.7 5.7 8.2 9.6 2.9
END
```

Chapter 3: Interpolation and Extrapolation

Chapter 3 of Numerical Recipes deals with interpolation and extrapolation (the same routines are usable for both). Three fundamental interpolation methods are first discussed,

1. *Polynomial interpolation (*`POLINT`*),*
2. *Rational function interpolation (*`RATINT`*), and*
3. *Cubic spline interpolation (*`SPLINE`*,* `SPLINT`*).*

To find the place in an ordered table at which to perform an interpolation, two routines are given, `LOCATE` *and* `HUNT`*. Also, for cases in which the actual coefficients of a polynomial interpolation are desired, the routines* `POLCOE` *and* `POLCOF` *are provided (along with important warnings circumscribing their usefulness).*

For higher-dimensional interpolations, Numerical Recipes treats only problems on a regularly spaced grid. Routine `POLIN2` *does a two-dimensional polynomial interpolation that aims at accuracy rather than smoothness. When smooth interpolation is desired, the methods shown in* `BCUCOF` *and* `BCUINT` *for bicubic interpolation are recommended. In the case of two-dimensional spline interpolations, the routines* `SPLIE2` *and* `SPLIN2` *are offered.*

⋆ ⋆ ⋆ ⋆

Program `POLINT` takes two arrays `XA` and `YA` of length N that express the known values of a function, and calculates the value, at a point x, of the unique polynomial of degree $N-1$ passing through all the given values. For the purpose of illustration, in `D3R1` we have taken evenly spaced `XA(I)` and set `YA(I)` equal to simple functions (sines and exponentials) of these `XA(I)`. For the sine we use an interval of length π, and for the exponential an interval of length 1.0. You may choose the number N of reference points and observe the improvement of the results as N increases. The test points x are slightly shifted from the reference points so that you can compare the estimated error `DY` with the actual error. By removing the shift, you may check that the polynomial actually hits all reference points.

```
      PROGRAM D3R1
C     Driver for routine POLINT
      PARAMETER(NP=10,PI=3.1415926)
      DIMENSION XA(NP),YA(NP)
      WRITE(*,*) 'Generation of interpolation tables'
      WRITE(*,*) ' ... sin(x)    0<x<pi'
      WRITE(*,*) ' ... exp(x)    0<x<1 '
```

```
      WRITE(*,*) 'How many entries go in these tables? (note: N<10)'
      READ(*,*) N
      DO 14 NFUNC=1,2
        IF (NFUNC.EQ.1) THEN
          WRITE(*,*) 'sine function from 0 to pi'
          DO 11 I=1,N
            XA(I)=I*PI/N
            YA(I)=SIN(XA(I))
11        CONTINUE
        ELSE IF (NFUNC.EQ.2) THEN
          WRITE(*,*) 'exponential function from 0 to 1'
          DO 12 I=1,N
            XA(I)=I*1.0/N
            YA(I)=EXP(XA(I))
12        CONTINUE
        ELSE
          STOP
        ENDIF
        WRITE(*,'(T10,A1,T20,A4,T28,A12,T46,A5)')
     *            'x','f(x)','interpolated','error'
        DO 13 I=1,10
          IF (NFUNC.EQ.1) THEN
            X=(-0.05+I/10.0)*PI
            F=SIN(X)
          ELSE IF (NFUNC.EQ.2) THEN
            X=(-0.05+I/10.0)
            F=EXP(X)
          ENDIF
          CALL POLINT(XA,YA,N,X,Y,DY)
          WRITE(*,'(1X,3F12.6,E15.4)') X,F,Y,DY
13      CONTINUE
        WRITE(*,*) '***********************************'
        WRITE(*,*) 'Press RETURN'
        READ(*,*)
14    CONTINUE
      END
```

RATINT is functionally similar to POLINT in that it also returns a value y for the function at point x, and an error estimate DY as well. In this case the values are determined from the unique diagonal rational function that passes through all the reference points. If you inspect the driver closely, you will find that two of the test points fall directly on top of reference points and should give exact results. The remainder do not. You can compare the estimated error DYY to the actual error |YY − YEXP| for these cases.

```
      PROGRAM D3R2
C     Driver for routine RATINT
      PARAMETER(NPT=6,EPSSQ=1.0)
      DIMENSION X(NPT),Y(NPT)
      F(X)=X*EXP(-X)/((X-1.0)**2+EPSSQ)
      DO 11 I=1,NPT
        X(I)=I*2.0/NPT
        Y(I)=F(X(I))
11    CONTINUE
      WRITE(*,'(/1X,A/)') 'Diagonal rational function interpolation'
      WRITE(*,'(1X,T6,A,T13,A,T26,A,T40,A)')
```

```
     *         'x','interp.','accuracy','actual'
        DO 12 I=1,10
          XX=0.2*I
          CALL RATINT(X,Y,NPT,XX,YY,DYY)
          YEXP=F(XX)
          WRITE(*,'(1X,F6.2,F12.6,E15.4,F12.6)') XX,YY,DYY,YEXP
12      CONTINUE
        END
```

Subroutine SPLINE generates a cubic spline. Given an array of x_i and $f(x_i)$, and given values of the first derivative of function f at the two endpoints of the tabulated region, it returns the second derivative of f at each of the tabulation points. As an example we chose the function $\sin x$ and evaluated it at evenly spaced points X(I). In this case the first derivatives at the end-points are YP1 $= \cos x_1$ and YPN $= \cos x_N$. The output array of SPLINE is Y2(I) and this is listed along with $-\sin x_i$, the second derivative of $\sin x_i$, for comparison.

```
        PROGRAM D3R3
C       Driver for routine SPLINE
        PARAMETER(N=20,PI=3.141593)
        DIMENSION X(N),Y(N),Y2(N)
        WRITE(*,*) 'Second-derivatives for sin(x) from 0 to PI'
C       Generate array for interpolation
        DO 11 I=1,20
          X(I)=I*PI/N
          Y(I)=SIN(X(I))
11      CONTINUE
C       Calculate 2nd derivative with SPLINE
        YP1=COS(X(1))
        YPN=COS(X(N))
        CALL SPLINE(X,Y,N,YP1,YPN,Y2)
C       Test result
        WRITE(*,'(T19,A,T35,A)') 'spline','actual'
        WRITE(*,'(T6,A,T17,A,T33,A)') 'angle','2nd deriv','2nd deriv'
        DO 12 I=1,N
          WRITE(*,'(1X,F8.2,2F16.6)') X(I),Y2(I),-SIN(X(I))
12      CONTINUE
        END
```

Actual cubic-spline interpolations, however, are carried out by SPLINT. This routine uses the output array from one call to SPLINE to service any subsequent number of spline interpolations with different x's. The demonstration program D3R4 tests this capability on both $\sin x$ and $\exp x$. The two are treated in succession according to whether NFUNC is one or two. In each case the function is tabulated at equally spaced points, and the derivatives are found at the first and last point. A call to SPLINE then produces an array of second derivatives Y2 which is fed to SPLINT. The interpolated values Y are compared with actual function values F at a different set of equally spaced points.

```
        PROGRAM D3R4
C       Driver for routine SPLINT, which calls SPLINE
        PARAMETER(NP=10,PI=3.141593)
        DIMENSION XA(NP),YA(NP),Y2(NP)
        DO 14 NFUNC=1,2
          IF (NFUNC.EQ.1) THEN
```

```
            WRITE(*,*) 'Sine function from 0 to pi'
            DO 11 I=1,NP
              XA(I)=I*PI/NP
              YA(I)=SIN(XA(I))
11          CONTINUE
            YP1=COS(XA(1))
            YPN=COS(XA(NP))
          ELSE IF (NFUNC.EQ.2) THEN
            WRITE(*,*) 'Exponential function from 0 to 1'
            DO 12 I=1,NP
              XA(I)=1.0*I/NP
              YA(I)=EXP(XA(I))
12          CONTINUE
            YP1=EXP(XA(1))
            YPN=EXP(XA(NP))
          ELSE
            STOP
          ENDIF
C       Call SPLINE to get second derivatives
          CALL SPLINE(XA,YA,NP,YP1,YP2,Y2)
C       Call SPLINT for interpolations
          WRITE(*,'(1X,T10,A1,T20,A4,T28,A13)') 'x','f(x)','interpolation'
          DO 13 I=1,10
            IF (NFUNC.EQ.1) THEN
              X=(-0.05+I/10.0)*PI
              F=SIN(X)
            ELSE IF (NFUNC.EQ.2) THEN
              X=-0.05+I/10.0
              F=EXP(X)
            ENDIF
            CALL SPLINT(XA,YA,Y2,NP,X,Y)
            WRITE(*,'(1X,3F12.6)') X,F,Y
13        CONTINUE
          WRITE(*,*) '***********************************'
          WRITE(*,*) 'Press RETURN'
          READ(*,*)
14      CONTINUE
        END
```

The next program, LOCATE, may be used in conjunction with any interpolation method to bracket the x-position for which $f(x)$ is sought by two adjacent tabulated positions. That is, given a monotonic array of x_i, and given a value of x, it finds the two values x_i, x_{i+1} that surround x. In D3R5 we chose the array x_i to be non-uniform, varying exponentially with i. Then we took a uniform series of x-values and sought their position in the array using LOCATE. For each X, LOCATE finds the value J for which X(J) is nearest below X. Then the driver shows J, and the two bracketing values XX(J) and XX(J+1). If J is 0 or N, then X is not within the tabulated range. The program thereby flags 'lower lim' if X is below X(1) or 'upper lim' if X is above X(N).

```
        PROGRAM D3R5
C       Driver for routine LOCATE
        PARAMETER(N=100)
        DIMENSION XX(N)
C       Create array to be searched
        DO 11 I=1,N
```

```
          XX(I)=EXP(I/20.0)-74.0
11      CONTINUE
        WRITE(*,*) 'Result of:   j=0 indicates x too small'
        WRITE(*,*) '            j=100 indicates x too large'
        WRITE(*,'(T5,A7,T17,A1,T24,A5,T34,A7)') 'locate ','j'
     *   ,'xx(j)','xx(j+1)'
C       Do test
        DO 12 I=1,19
          X=-100.0+200.0*I/20.0
          CALL LOCATE(XX,N,X,J)
          IF (J.EQ.0) THEN
            WRITE(*,'(1X,F10.4,I6,A12,F12.6)') X,J,'lower lim',XX(J+1)
          ELSE IF (J.EQ.N) THEN
            WRITE(*,'(1X,F10.4,I6,F12.6,A12)') X,J,XX(J),'upper lim'
          ELSE
            WRITE(*,'(1X,F10.4,I6,2F12.6)') X,J,XX(J),XX(J+1)
          ENDIF
12      CONTINUE
        END
```

Routine HUNT serves the same function as LOCATE, but is used when the table is to be searched many times and the abscissa each time is close to its value on the previous search. D3R6 sets up the array XX(I) and then a series X of points to locate. The hunt begins with a trial value JI (which is fed to HUNT through variable J) and HUNT returns solution J such that X lies between XX(J) and XX(J+1). The two cases J=0 and J=N have the same meaning as in D3R5 and are treated in the same way.

```
        PROGRAM D3R6
C       Driver for routine HUNT
        PARAMETER(N=100)
        DIMENSION XX(N)
C       Create array to be searched
        DO 11 I=1,N
          XX(I)=EXP(I/20.0)-74.0
11      CONTINUE
        WRITE(*,*) 'Result of:    j=0 indicates x too small'
        WRITE(*,*) '             j=100 indicates x too large'
        WRITE(*,'(T7,A7,T17,A5,T25,A1,T32,A5,T42,A7)') 'locate:',
     *   'guess','j','xx(j)','xx(j+1)'
C       Do test
        DO 12 I=1,19
          X=-100.0+200.0*I/20.0
C       Trial parameter
          JI=5*I
          J=JI
C       Begin search
          CALL HUNT(XX,N,X,J)
          IF (J.EQ.0) THEN
            WRITE(*,'(1X,F12.6,2I6,A12,F12.6)') X,JI,J,
     *              'lower lim',XX(J+1)
          ELSE IF (J.EQ.N) THEN
            WRITE(*,'(1X,F12.6,2I6,F12.6,A12)') X,JI,J,
     *              XX(J),'upper lim'
          ELSE
            WRITE(*,'(1X,F12.6,2I6,2F12.6)') X,JI,J,XX(J),XX(J+1)
```

```
          ENDIF
12      CONTINUE
        END
```

The next two demonstration programs, D3R7 and D3R8, are so nearly identical that they may be discussed together. POLCOE and POLCOF themselves both find coefficients of interpolating polynomials. In the present instance we have tried both a sine function and an exponential function for YA(I), each tabulated at uniformly spaced points XA(I). The validity of the array of polynomial coefficients COEFF is tested by calculating the value SUM of the polynomials at a series of test points and listing these alongside the functions F which they represent.

```
        PROGRAM D3R7
C       Driver for routine POLCOE
        PARAMETER(NP=5,PI=3.1415926)
        DIMENSION XA(NP),YA(NP),COEFF(NP)
        DO 15 NFUNC=1,2
          IF (NFUNC.EQ.1) THEN
            WRITE(*,*) 'Sine function from 0 to PI'
            DO 11 I=1,NP
              XA(I)=I*PI/NP
              YA(I)=SIN(XA(I))
11          CONTINUE
          ELSE IF (NFUNC.EQ.2) THEN
            WRITE(*,*) 'Exponential function from 0 to 1'
            DO 12 I=1,NP
              XA(I)=1.0*I/NP
              YA(I)=EXP(XA(I))
12          CONTINUE
          ELSE
            STOP
          ENDIF
          CALL POLCOE(XA,YA,NP,COEFF)
          WRITE(*,*) '    coefficients'
          WRITE(*,'(1X,6F12.6)') (COEFF(I),I=1,NP)
          WRITE(*,'(1X,T10,A1,T20,A4,T29,A10)')
     *          'x','f(x)','polynomial'
          DO 14 I=1,10
            IF (NFUNC.EQ.1) THEN
              X=(-0.05+I/10.0)*PI
              F=SIN(X)
            ELSE IF (NFUNC.EQ.2) THEN
              X=-0.05+I/10.0
              F=EXP(X)
            ENDIF
            SUM=COEFF(NP)
            DO 13 J=NP-1,1,-1
              SUM=COEFF(J)+SUM*X
13          CONTINUE
            WRITE(*,'(1X,3F12.6)') X,F,SUM
14        CONTINUE
          WRITE(*,*) '***********************************'
          WRITE(*,*) 'Press RETURN'
          READ(*,*)
15      CONTINUE
        END
```

```
      PROGRAM D3R8
C     Driver for routine POLCOF
      PARAMETER(NP=5,PI=3.141593)
      DIMENSION XA(NP),YA(NP),COEFF(NP)
      DO 15 NFUNC=1,2
        IF (NFUNC.EQ.1) THEN
          WRITE(*,*) 'Sine function from 0 to PI'
          DO 11 I=1,NP
            XA(I)=I*PI/NP
            YA(I)=SIN(XA(I))
11        CONTINUE
        ELSE IF (NFUNC.EQ.2) THEN
          WRITE(*,*) 'Exponential function from 0 to 1'
          DO 12 I=1,NP
            XA(I)=1.0*I/NP
            YA(I)=EXP(XA(I))
12        CONTINUE
        ELSE
          STOP
        ENDIF
        CALL POLCOF(XA,YA,NP,COEFF)
        WRITE(*,*) '    coefficients'
        WRITE(*,'(1X,6F12.6)') (COEFF(I),I=1,NP)
        WRITE(*,'(1X,T10,A1,T20,A4,T29,A10)')
     *          'x','f(x)','polynomial'
        DO 14 I=1,10
          IF (NFUNC.EQ.1) THEN
            X=(-0.05+I/10.0)*PI
            F=SIN(X)
          ELSE IF (NFUNC.EQ.2) THEN
            X=-0.05+I/10.0
            F=EXP(X)
          ENDIF
          SUM=COEFF(NP)
          DO 13 J=NP-1,1,-1
            SUM=COEFF(J)+SUM*X
13        CONTINUE
          WRITE(*,'(1X,3F12.6)') X,F,SUM
14      CONTINUE
        WRITE(*,*) '***********************************'
        WRITE(*,*) 'Press RETURN'
        READ(*,*)
15    CONTINUE
      END
```

For two-dimensional interpolation, POLIN2 implements a bilinear interpolation. We feed it coordinates X1A,X2A for an $M \times N$ array of gridpoints as well as the function value at each gridpoint. In return it gives the value Y of the interpolated function at a given point X1,X2, and the estimated accuracy DY of the interpolation. D3R9 runs the test on a uniform grid for the function $f(x,y) = \sin x \exp y$. Then, for an offset grid of test points, the interpolated value Y is compared to the actual function value F, and the actual error is compared to the estimated error DY.

```
      PROGRAM D3R9
C     Driver for routine POLIN2
      PARAMETER(N=5,PI=3.141593)
```

```
      DIMENSION X1A(N),X2A(N),YA(N,N)
      DO 12 I=1,N
          X1A(I)=I*PI/N
          DO 11 J=1,N
              X2A(J)=1.0*J/N
              YA(I,J)=SIN(X1A(I))*EXP(X2A(J))
11        CONTINUE
12        CONTINUE
C     Test 2-dimensional interpolation
      WRITE(*,'(T9,A,T21,A,T32,A,T40,A,T58,A)')
     *  'x1','x2','f(x)','interpolated','error'
      DO 14 I=1,4
      X1=(-0.1+I/5.0)*PI
          DO 13 J=1,4
              X2=-0.1+J/5.0
              F=SIN(X1)*EXP(X2)
              CALL POLIN2(X1A,X2A,YA,N,N,X1,X2,Y,DY)
              WRITE(*,'(1X,4F12.6,F14.6)') X1,X2,F,Y,DY
13        CONTINUE
          WRITE(*,*) '***********************************'
14    CONTINUE
      END
```

Bicubic interpolation in two dimensions is carried out with BCUCOF and BCUINT. The first supplies interpolating coefficients within a grid square and the second calculates interpolated values. The calculation provides not only interpolated function values, but also interpolated values of two partial derivatives, all of which are guaranteed to be smooth. To get this, we are required to supply more information than we have needed in previous interpolation routines.

Demonstration program D3R10 works with the function $f(x,y) = xy\exp(-xy)$. You may compare the two first derivatives and the cross derivative of this function with what you find computed in the routine. The function and derivatives are calculated at the four corners of a rectangular grid cell, in this case a 2×2 unit square with one corner at the origin. The points are supplied counterclockwise around the cell. D1 and D2 are the dimensions of the cell. A call to BCUCOF provides sixteen coefficients which are listed below for your reference.

```
 Coefficients for bicubic interpolation
    0.000000E+00   0.000000E+00   0.000000E+00   0.000000E+00
    0.000000E+00   0.400000E+01   0.000000E+00   0.000000E+00
    0.000000E+00   0.000000E+00  -0.136556E+02   0.609517E+01
    0.000000E+00   0.000000E+00   0.609517E+01  -0.246149E+01
```

```
      PROGRAM D3R10
C     Driver for routine BCUCOF
      DIMENSION C(4,4),Y(4),Y1(4),Y2(4)
      DIMENSION Y12(4),X1(4),X2(4)
      DATA X1/0.0,2.0,2.0,0.0/
      DATA X2/0.0,0.0,2.0,2.0/
      D1=X1(2)-X1(1)
      D2=X2(4)-X2(1)
      DO 11 I=1,4
          X1X2=X1(I)*X2(I)
          EE=EXP(-X1X2)
          Y(I)=X1X2*EE
```

```
          Y1(I)=X2(I)*(1.0-X1X2)*EE
          Y2(I)=X1(I)*(1.0-X1X2)*EE
          Y12(I)=(1.0-3.0*X1X2+X1X2**2)*EE
11      CONTINUE
        CALL BCUCOF(Y,Y1,Y2,Y12,D1,D2,C)
        WRITE(*,*) 'Coefficients for bicubic interpolation'
        DO 12 I=1,4
          WRITE(*,'(1X,4e15.6)') (C(I,J),J=1,4)
12      CONTINUE
        END
```

Program D3R11 works with the function $f(x, y) = (xy)^2$, which has derivatives $\partial f/\partial x = 2xy^2$, $\partial f/\partial y = 2yx^2$, and $\partial^2 f/\partial x \partial y = 4xy$. These are supplied to BCUINT along with the locations of the grid points. BCUINT calls BCUCOF internally to determine coefficients, and then calculates ANSY, ANSY1, ANSY2, the interpolated values of f, $\partial f/\partial x$ and $\partial f/\partial y$ at the specified test point (X1,X2). These are compared by the demonstration program to expected values for the three quantities, which are called EY, EY1, and EY2. The test points run along the diagonal of the grid square.

```
        PROGRAM D3R11
C       Driver for routine BCUINT
        DIMENSION Y(4),Y1(4),Y2(4),Y12(4),XX(4),YY(4)
        DATA XX/0.0,2.0,2.0,0.0/
        DATA YY/0.0,0.0,2.0,2.0/
        X1L=XX(1)
        X1U=XX(2)
        X2L=YY(1)
        X2U=YY(4)
        DO 11 I=1,4
          XXYY=XX(I)*YY(I)
          Y(I)=XXYY**2
          Y1(I)=2.0*YY(I)*XXYY
          Y2(I)=2.0*XX(I)*XXYY
          Y12(I)=4.0*XXYY
11      CONTINUE
        WRITE(*,'(/1X,T6,A,T14,A,T22,A,T28,A,T38,A,T44,A,T54,A,T60,A/)')
     *          'X1','X2','Y','EXPECT','Y1','EXPECT','Y2','EXPECT'
        DO 12 I=1,10
          X1=0.2*I
          X2=X1
          CALL BCUINT(Y,Y1,Y2,Y12,X1L,X1U,X2L,X2U,X1,X2,ANSY,ANSY1,ANSY2)
          X1X2=X1*X2
          EY=X1X2**2
          EY1=2.0*X2*X1X2
          EY2=2.0*X1*X1X2
          WRITE(*,'(1X,8F8.4)') X1,X2,ANSY,EY,ANSY1,EY1,ANSY2,EY2
12      CONTINUE
        END
```

Routines SPLIE2 and SPLIN2 work as a pair to perform bicubic spline interpolations. SPLIE2 takes a function tabulated on an $M \times N$ grid and performs one dimensional natural cubic splines along the rows of the grid to generate an array of second derivatives. These are fodder for SPLIN2 which takes the grid points, function values, and second derivative values and returns the interpolated function value for a desired point in the grid region.

Demonstration program D3R12 exercises SPLIE2 on a regular 10×10 grid of points with coordinates X1 and X2, for the function $y = (x_1 x_2)^2$. The calculated second derivative array is compared with the actual second derivative $2x_1x_2$ of the function. Keep in mind that a natural spline is assumed, so that agreement will not be so good near the boundaries of the grid. (This shows that you should *not* assume a natural spline if you have better derivative information at the endpoints.)

```
      PROGRAM D3R12
C     Driver for routine SPLIE2
      PARAMETER(M=10,N=10)
      DIMENSION X1(M),X2(N),Y(M,N),Y2(M,N)
      DO 11 I=1,M
        X1(I)=0.2*I
11    CONTINUE
      DO 12 I=1,N
        X2(I)=0.2*I
12    CONTINUE
      DO 14 I=1,M
        DO 13 J=1,N
          X1X2=X1(I)*X2(J)
          Y(I,J)=X1X2**2
13      CONTINUE
14    CONTINUE
      CALL SPLIE2(X1,X2,Y,M,N,Y2)
      WRITE(*,'(/1X,A)') 'Second derivatives from SPLIE2'
      WRITE(*,'(1X,A/)') 'Natural spline assumed'
      DO 15 I=1,5
        WRITE(*,'(1X,5F12.6)') (Y2(I,J),J=1,5)
15    CONTINUE
      WRITE(*,'(/1X,A/)') 'Actual second derivatives'
      DO 17 I=1,5
        DO 16 J=1,5
          Y2(I,J)=2.0*(X1(I)**2)
16      CONTINUE
        WRITE(*,'(1X,5F12.6)') (Y2(I,J),J=1,5)
17    CONTINUE
      END
```

The demonstration program D3R13 establishes a similar 10×10 grid for the function $y = x_1 x_2 \exp(-x_1 x_2)$. It makes a single call to SPLIE2 to produce second derivatives Y2, and then finds function values F through calls to SPLIN2, comparing them to actual function values FF. These values are determined, for no reason better than perversity, along a quadratic path $x_2 = x_1^2$ through the grid region.

```
      PROGRAM D3R13
C     Driver for routine SPLIN2
      PARAMETER(M=10,N=10)
      DIMENSION X1(M),X2(N),Y(M,N),Y2(M,N)
      DO 11 I=1,M
        X1(I)=0.2*I
11    CONTINUE
      DO 12 I=1,N
        X2(I)=0.2*I
12    CONTINUE
      DO 14 I=1,M
```

```
          DO 13 J=1,N
            X1X2=X1(I)*X2(J)
            Y(I,J)=X1X2*EXP(-X1X2)
13        CONTINUE
14      CONTINUE
        CALL SPLIE2(X1,X2,Y,M,N,Y2)
        WRITE(*,'(/1X,T9,A,T21,A,T31,A,T43,A)')
     *        'x1','x2','splin2','actual'
        DO 15 I=1,10
          XX1=0.1*I
          XX2=XX1**2
          CALL SPLIN2(X1,X2,Y,Y2,M,N,XX1,XX2,F)
          X1X2=XX1*XX2
          FF=X1X2*EXP(-X1X2)
          WRITE(*,'(1X,4F12.6)') XX1,XX2,F,FF
15      CONTINUE
        END
```

Chapter 4: Integration of Functions

Numerical integration, or "quadrature", has been treated with some degree of detail in Numerical Recipes . Chapter 4 begins with TRAPZD, *a subroutine for applying the extended trapezoidal rule. It can be used in successive calls for sequentially improving accuracy, and is used as a foundation for several other programs. For example* QTRAP *is an integrating routine that makes repeated calls to* TRAPZD *until a certain fractional accuracy is achieved.* QSIMP *also calls* TRAPZD, *and in this case performs integration by Simpson's rule. Romberg integration, a generalization of Simpson's rule to successively higher orders, is performed with* QROMB*—this one also calls* TRAPZD. *For improper integrals a different "workhorse" is used, the subroutine* MIDPNT. *This routine applies the extended midpoint rule to avoid function evaluations at an endpoint of the region of integration. It can be used in* QTRAP *or* QSIMP *in place of* TRAPZD. *Routine* QROMB *can be generalized similarly, and we have implemented this idea in* QROMO, *a Romberg integrator for open intervals. The chapter also offers a number of exact replacements for* MIDPNT, *to be used for various types of singularity in the integrand:*

1. MIDINF *– if one or the other of the limits of integration is infinite.*
2. MIDSQL *- if there is an inverse square root singularity of the integrand at the lower limit of integration.*
3. MIDSQU *- if there is an inverse square root singularity of the integrand at the upper limit of integration.*
4. MIDEXP *- when the upper limit of integration is infinite and the integrand decreases exponentially at infinity.*

The somewhat more subtle method of Gaussian quadrature uses unequally spaced abscissas, and weighting coefficients which can be read from tables. Routine QGAUS *computes integrals with a ten-point Gauss-Legendre weighting using such coefficients.* GAULEG *calculates the tables of abscissas and weights that would apply to an N-point Gauss-Legendre quadrature.*

⋆ ⋆ ⋆ ⋆

TRAPZD applies the extended trapezoidal rule for integration. It is called sequentially for higher and higher stages of refinement of the integral. The sample program D4R1 uses TRAPZD to perform a numerical integration of the function

$$\text{FUNC} = x^2(x^2 - 2)\sin x$$

whose definite integral is

$$\text{FINT} = 4x(x^2 - 7)\sin x - (x^4 - 14x^2 + 28)\cos x.$$

The integral is performed from $A = 0.0$ to $B = \pi/2$. To demonstrate the increasing accuracy on sequential calls, TRAPZD is called 14 times with the index I increasing by one each time. The improving values of the integral are listed for comparison to the actual value FINT(B) − FINT(A).

```
      PROGRAM D4R1
C     Driver for routine TRAPZD
      PARAMETER(NMAX=14, PIO2=1.5707963)
      EXTERNAL FUNC,FINT
      A=0.0
      B=PIO2
      WRITE(*,'(1X,A)') 'Integral of FUNC with 2^(n-1) points'
      WRITE(*,'(1X,A,F10.6)') 'Actual value of integral is',FINT(B)-FINT(A)
      WRITE(*,'(1X,T7,A,T16,A)') 'n','Approx. Integral'
      DO 11 I=1,NMAX
          CALL TRAPZD(FUNC,A,B,S,I)
          WRITE(*,'(1X,I6,F20.6)') I,S
11    CONTINUE
      END
      FUNCTION FUNC(X)
      FUNC=(X**2)*(X**2-2.0)*SIN(X)
      END
      FUNCTION FINT(X)
C     Integral of FUNC
      FINT=4.0*X*((X**2)-7.0)*SIN(X)
     *        -((X**4)-14.0*(X**2)+28.0)*COS(X)
      END
```

QTRAP carries out the same integration algorithm but allows us to specify the accuracy with which we wish the integration done. (It is specified within QTRAP as EPS=1.0E-6.) QTRAP itself makes the sequential calls to TRAPZD until the desired accuracy is reached. Then QTRAP issues a single result. In sample program D4R2 we compare this result to the exact value of the integral.

```
      PROGRAM D4R2
C     Driver for routine QTRAP
      PARAMETER(PIO2=1.5707963)
      EXTERNAL FUNC,FINT
      A=0.0
      B=PIO2
      WRITE(*,'(1X,A)') 'Integral of FUNC computed with QTRAP'
      WRITE(*,'(1X,A,F10.6)') 'Actual value of integral is',
     *        FINT(B)-FINT(A)
      CALL QTRAP(FUNC,A,B,S)
      WRITE(*,'(1X,A,F10.6)') 'Result from routine QTRAP is',S
      END
      FUNCTION FUNC(X)
      FUNC=(X**2)*(X**2-2.0)*SIN(X)
      END
      FUNCTION FINT(X)
C     Integral of FUNC
      FINT=4.0*X*((X**2)-7.0)*SIN(X)
     *        -((X**4)-14.0*(X**2)+28.0)*COS(X)
```

```
      END
```

Alternatively, the integral may be handled by QSIMP which applies Simpson's rule. Sample program D4R3 carries out the same integration as the previous program, and reports the result in the same way as well.

```
      PROGRAM D4R3
C     Driver for routine QSIMP
      PARAMETER(PIO2=1.5707963)
      EXTERNAL FUNC,FINT
      A=0.0
      B=PIO2
      WRITE(*,'(1X,A)') 'Integral of FUNC computed with QSIMP'
      WRITE(*,'(1X,A,F10.6)') 'Actual value of integral is',
     *        FINT(B)-FINT(A)
      CALL QSIMP(FUNC,A,B,S)
      WRITE(*,'(1X,A,F10.6)') 'Result from routine QSIMP is',S
      END
      FUNCTION FUNC(X)
      FUNC=(X**2)*(X**2-2.0)*SIN(X)
      END
      FUNCTION FINT(X)
C     Integral of FUNC
      FINT=4.0*X*((X**2)-7.0)*SIN(X)
     *        -((X**4)-14.0*(X**2)+28.0)*COS(X)
      END
```

QROMB generalizes Simpson's rule to higher orders. It makes successive calls to TRAPZD and stores the results. Then it uses POLINT, the polynomial interpolater/extrapolator, to project the value of integral which would be obtained were we to continue indefinitely with TRAPZD. Sample program D4R4 is essentially identical to the sample programs for QTRAP and QSIMP.

```
      PROGRAM D4R4
C     Driver for routine QROMB
      PARAMETER(PIO2=1.5707963)
      EXTERNAL FUNC,FINT
      A=0.0
      B=PIO2
      WRITE(*,'(1X,A)') 'Integral of FUNC computed with QROMB'
      WRITE(*,'(1X,A,F10.6)') 'Actual value of integral is',
     *        FINT(B)-FINT(A)
      CALL QROMB(FUNC,A,B,S)
      WRITE(*,'(1X,A,F10.6)') 'Result from routine QROMB is',S
      END
      FUNCTION FUNC(X)
      FUNC=(X**2)*(X**2-2.0)*SIN(X)
      END
      FUNCTION FINT(X)
C     Integral of FUNC
      FINT=4.0*X*((X**2)-7.0)*SIN(X)
     *        -((X**4)-14.0*(X**2)+28.0)*COS(X)
      END
```

Sample program D4R5 uses the function FUNC2 $= 1/\sqrt{x}$ which is singular at the origin. Limits of integration are set at $A = 0.0$ and $B = 1.0$. MIDPNT, however, implements an open formula and does not evaluate the function exactly at $x = 0$. In this case the integral is compared to FINT2(B) − FINT2(A) where FINT2 = $2\sqrt{x}$, the integral of FUNC2.

```
      PROGRAM D4R5
C     Driver for routine MIDPNT
      PARAMETER(NMAX=10)
      EXTERNAL FUNC2,FINT2
      A=0.0
      B=1.0
      WRITE(*,*) 'Integral of FUNC2 computed with MIDPNT'
      WRITE(*,*) 'Actual value of integral is',FINT2(B)-FINT2(A)
      WRITE(*,'(1X,T7,A,T20,A)') 'n','Approx. Integral'
      DO 11 I=1,NMAX
        CALL MIDPNT(FUNC2,A,B,S,I)
        WRITE(*,'(1X,I6,F24.6)') I,S
11    CONTINUE
      END
      FUNCTION FUNC2(X)
      FUNC2=1.0/SQRT(X)
      END
      FUNCTION FINT2(X)
C     Integral of FUNC2
      FINT2=2.0*SQRT(X)
      END
```

Various special forms of MIDPNT (i.e. MIDSQL, MIDSQU, MIDINF) are demonstrated by sample program D4R6. For those tests that integrate to infinity, we take infinity to be 1.0E20. The following integrations are performed:

1. Integral of $\sqrt{x}/\sin x$ from 0.0 to $\pi/2$. (This has a $1/\sqrt{x}$ singularity at $x = 0$, and uses MIDSQL.)
2. Integral of $\sqrt{(\pi - x)}/\sin x$ from $\pi/2$ to π. (This has the $1/\sqrt{x}$ singularity at the upper limit $x = \pi$, and uses MIDSQU.)
3. Integral of $(\sin x)/x^2$ from $\pi/2$ to ∞. (This has a region of integration extending to ∞ and uses MIDINF. It is quite slowly convergent, as is the next integral.)
4. Integral of $(\sin x)/x^2$ from $-\infty$ to $-\pi/2$. (Region of integration goes to $-\infty$; uses MIDINF.)
5. Integral of $\exp(-x)/\sqrt{x}$ from 0.0 to ∞. (This has a singularity at $x = 0.0$ and also integrates to ∞. It is performed in two pieces, (0.0 to $\pi/2$) and ($\pi/2$ to ∞) using MIDSQL and MIDINF. The two calculations give results RES1 and RES2 respectively, which are added to give the entire integral.)

```
      PROGRAM D4R6
C     Driver for QROMO
      PARAMETER(X1=0.0,X2=1.5707963,X3=3.1415926,AINF=1.0E20)
      EXTERNAL FUNCL,MIDSQL,FUNCU,MIDSQU,FNCINF,MIDINF,FNCEND
      WRITE(*,'(/1X,A)') 'Improper integrals:'
      CALL QROMO(FUNCL,X1,X2,RESULT,MIDSQL)
      WRITE(*,'(/1X,A)')
     *        'Function: SQRT(x)/SIN(x)       Interval: (0,pi/2)'
      WRITE(*,'(1X,A,F8.4)')
     *        'Using: MIDSQL                  Result:',RESULT
```

```
      CALL QROMO(FUNCU,X2,X3,RESULT,MIDSQU)
      WRITE(*,'(/1X,A)')
     *        'Function: SQRT(pi-x)/SIN(x)     Interval: (pi/2,pi)'
      WRITE(*,'(1X,A,F8.4)')
     *        'Using: MIDSQU                   Result:',RESULT
      CALL QROMO(FNCINF,X2,AINF,RESULT,MIDINF)
      WRITE(*,'(/1X,A)')
     *        'Function: SIN(x)/x**2           Interval: (pi/2,infty)'
      WRITE(*,'(1X,A,F8.4)')
     *        'Using: MIDINF                   Result:',RESULT
      CALL QROMO(FNCINF,-AINF,-X2,RESULT,MIDINF)
      WRITE(*,'(/1X,A)')
     *        'Function: SIN(x)/x**2           Interval: (-infty,-pi/2)'
      WRITE(*,'(1X,A,F8.4)')
     *        'Using: MIDINF                   Result:',RESULT
      CALL QROMO(FNCEND,X1,X2,RES1,MIDSQL)
      CALL QROMO(FNCEND,X2,AINF,RES2,MIDINF)
      WRITE(*,'(/1X,A)')
     *        'Function: EXP(-x)/SQRT(x)       Interval: (0.0,infty)'
      WRITE(*,'(1X,A,F8.4/)')
     *        'Using: MIDSQL,MIDINF            Result:',RES1+RES2
      END
      SUBROUTINE MIDSQL(FUNK,AA,BB,S,N)
      FUNC(X)=2.*X*FUNK(AA+X**2)
      B=SQRT(BB-AA)
      A=0.0
      IF (N.EQ.1) THEN
          S=(B-A)*FUNC(0.5*(A+B))
          IT=1
      ELSE
          TNM=IT
          DEL=(B-A)/(3.*TNM)
          DDEL=DEL+DEL
          X=A+0.5*DEL
          SUM=0.
          DO 11 J=1,IT
              SUM=SUM+FUNC(X)
              X=X+DDEL
              SUM=SUM+FUNC(X)
              X=X+DEL
11        CONTINUE
          S=(S+(B-A)*SUM/TNM)/3.
          IT=3*IT
      ENDIF
      RETURN
      END
      SUBROUTINE MIDSQU(FUNK,AA,BB,S,N)
      FUNC(X)=2.0*X*FUNK(BB-X**2)
      B=SQRT(BB-AA)
      A=0.0
      IF (N.EQ.1) THEN
          S=(B-A)*FUNC(0.5*(A+B))
          IT=1
      ELSE
          TNM=IT
          DEL=(B-A)/(3.*TNM)
          DDEL=DEL+DEL
```

```
            X=A+0.5*DEL
            SUM=0.
            DO 11 J=1,IT
                SUM=SUM+FUNC(X)
                X=X+DDEL
                SUM=SUM+FUNC(X)
                X=X+DEL
11          CONTINUE
            S=(S+(B-A)*SUM/TNM)/3.
            IT=3*IT
        ENDIF
        RETURN
        END
        FUNCTION FUNCL(X)
        FUNCL=SQRT(X)/SIN(X)
        END
        FUNCTION FUNCU(X)
        PARAMETER(PI=3.1415926)
        FUNCU=SQRT(PI-X)/SIN(X)
        END
        FUNCTION FNCINF(X)
        FNCINF=SIN(X)/(X**2)
        END
        FUNCTION FNCEND(X)
        FNCEND=EXP(-X)/SQRT(X)
        END
```

Subroutine QGAUS performs a Gauss-Legendre integration, using only ten function evaluations. Sample program D4R8 applies it to the function $x\exp(-x)$ whose integral from x_1 to x is $(1+x_1)\exp(-x_1) - (1+x)\exp(-x)$. QGAUS returns this integral as parameter SS. The method is used for a series of intervals, as short as $(0.0 - 0.5)$ and as long as $(0.0 - 5.0)$. You may observe how the accuracy depends on the interval.

```
        PROGRAM D4R8
C       Driver for routine QGAUS
        EXTERNAL FUNC
        PARAMETER(X1=0.0,X2=5.0,NVAL=10)
        DX=(X2-X1)/NVAL
        WRITE(*,'(/1X,A,T12,A,T23,A/)') '0.0 to','QGAUS','Expected'
        DO 11 I=1,NVAL
            X=X1+I*DX
            CALL QGAUS(FUNC,X1,X,SS)
            WRITE(*,'(1X,F5.2,2F12.6)') X,SS,
     *              -(1.0+X)*EXP(-X)+(1.0+X1)*EXP(-X1)
11      CONTINUE
        END
        FUNCTION FUNC(X)
        FUNC=X*EXP(-X)
        END
```

Sample program D4R9, which drives GAULEG, performs the same method of quadrature, and on the same function. However, it chooses its own abscissas and weights for the Gauss-Legendre calculation, and is not restricted to a ten-point formula; it can do an N-point calculation for any N. The N abscissas and weights appropriate to an interval $x = 0.0$ to 1.0 are found by sample program D4R9 for the case $N = 10$. The results you should find are listed below. Next the program applies these values to a quadrature and compares the result to that from a formal integration.

```
position        weight
.013047        .033336
.067468        .074726
.160295        .109543
.283302        .134633
.425563        .147762
.574437        .147762
.716698        .134633
.839705        .109543
.932532        .074726
.986953        .033336
```

```
      PROGRAM D4R9
C     Driver for routine GAULEG
      PARAMETER(NPOINT=10,X1=0.0,X2=1.0,X3=10.0)
      DIMENSION X(NPOINT),W(NPOINT)
      CALL GAULEG(X1,X2,X,W,NPOINT)
      WRITE(*,'(/1X,T3,A,T10,A,T22,A/)') '#','X(I)','W(I)'
      DO 11 I=1,NPOINT
        WRITE(*,'(1X,I2,2F12.6)') I,X(I),W(I)
11    CONTINUE
C     Demonstrate the use of GAULEG for an integral
      CALL GAULEG(X1,X3,X,W,NPOINT)
      XX=0.0
      DO 12 I=1,NPOINT
        XX=XX+W(I)*FUNC(X(I))
12    CONTINUE
      WRITE(*,'(/1X,A,F12.6)') 'Integral from GAULEG:',XX
      WRITE(*,'(1X,A,F12.6)') 'Actual value:',1.0-(1.0+X3)*EXP(-X3)
      END
      FUNCTION FUNC(X)
      FUNC=X*EXP(-X)
      END
```

Chapter 4 of *Numerical Recipes* ends with a short discussion of multidimensional integration, exemplified by routine QUAD3D which does a 3-dimensional integration by repeated 1-dimensional integration. Sample program D4R10 applies the method to the integration of FUNC $= x^2 + y^2 + z^2$ over a spherical volume with a radius XMAX which is taken successively as $0.1, 0.2, ..., 1.0$. The integral is done in Cartesian rather than spherical coordinates, but the result is compared to that easily found in spherical coordinates, $4\pi(\text{XMAX})^5/5$. The concept is quite simple, but you may find the zoo of subroutines somewhat confusing. Subroutine FUNC generates the function. Subroutines Y1 and Y2 supply the two limits of the y-integration for each value of x. Similarly Z1 and Z2 give the limits of z-integration for given x and y. Routines QGAUSX, QGAUSY, and QGAUSZ are all identical except for their names, and are used to do Gauss-Legendre integration along the three coordinate directions.

```
      PROGRAM D4R10
C     Driver for routine QUAD3D
      COMMON XMAX
      PARAMETER(PI=3.1415926,NVAL=10)
      WRITE(*,*) 'Integral of r^2 over a spherical volume'
```

```
      WRITE(*,'(/4X,A,T14,A,T24,A)') 'Radius','QUAD3D','Actual'
      DO 11 I=1,NVAL
        XMAX=0.1*I
        XMIN=-XMAX
        CALL QUAD3D(XMIN,XMAX,S)
        WRITE(*,'(1X,F8.2,2F10.4)') XMAX,S,4.0*PI*(XMAX**5)/5.0
11    CONTINUE
      END
      FUNCTION FUNC(X,Y,Z)
      FUNC=X**2+Y**2+Z**2
      END
      FUNCTION Z1(X,Y)
      COMMON XMAX
      Z1=-SQRT(ABS(XMAX**2-X**2-Y**2))
      END
      FUNCTION Z2(X,Y)
      COMMON XMAX
      Z2=SQRT(ABS(XMAX**2-X**2-Y**2))
      END
      FUNCTION Y1(X)
      COMMON XMAX
      Y1=-SQRT(ABS(XMAX**2-X**2))
      END
      FUNCTION Y2(X)
      COMMON XMAX
      Y2=SQRT(ABS(XMAX**2-X**2))
      END
      SUBROUTINE QGAUSX(FUNC,A,B,SS)
      DIMENSION X(5),W(5)
      DATA X/.1488743389,.4333953941,.6794095682,
     *        .8650633666,.9739065285/
      DATA W/.2955242247,.2692667193,.2190863625,
     *        .1494513491,.0666713443/
      XM=0.5*(B+A)
      XR=0.5*(B-A)
      SS=0
      DO 11 J=1,5
        DX=XR*X(J)
        SS=SS+W(J)*(FUNC(XM+DX)+FUNC(XM-DX))
11    CONTINUE
      SS=XR*SS
      RETURN
      END
      SUBROUTINE QGAUSY(FUNC,A,B,SS)
      DIMENSION X(5),W(5)
      DATA X/.1488743389,.4333953941,.6794095682,
     *        .8650633666,.9739065285/
      DATA W/.2955242247,.2692667193,.2190863625,
     *        .1494513491,.0666713443/
      XM=0.5*(B+A)
      XR=0.5*(B-A)
      SS=0
      DO 11 J=1,5
        DX=XR*X(J)
        SS=SS+W(J)*(FUNC(XM+DX)+FUNC(XM-DX))
11    CONTINUE
      SS=XR*SS
```

```
      RETURN
      END
      SUBROUTINE QGAUSZ(FUNC,A,B,SS)
      DIMENSION X(5),W(5)
      DATA X/.1488743389,.4333953941,.6794095682,
     *        .8650633666,.9739065285/
      DATA W/.2955242247,.2692667193,.2190863625,
     *        .1494513491,.0666713443/
      XM=0.5*(B+A)
      XR=0.5*(B-A)
      SS=0
      DO 11 J=1,5
        DX=XR*X(J)
        SS=SS+W(J)*(FUNC(XM+DX)+FUNC(XM-DX))
11    CONTINUE
      SS=XR*SS
      RETURN
      END
```

Chapter 5: Evaluation of Functions

Chapter 5 of Numerical Recipes treats the approximation and evaluation of functions. The methods, along with a few others, are applied in Chapter 6 to the calculation of a collection of "special" functions. Polynomial or power series expansions are perhaps the most often used approximations and a few tips are given for accelerating the convergence of some series. In the case of alternating series, Euler's transformation is popular, and is implemented in program EULSUM. *For general polynomials,* DDPOLY *demonstrates the evaluation of both the polynomial and its derivatives from a list of its coefficients. The division of one polynomial into another, giving a quotient and remainder polynomial, is done by* POLDIV.

The approximation of functions by Chebyshev polynomial series is presented as a method of arriving at the approximation of nearly smallest deviation from the true function over a given region for a specified order of approximation. The coefficients for such polynomials are given by CHEBFT *and function approximations are subsequently carried out by* CHEBEV. *To generate the derivative or integral of a function from its Chebyshev coefficients, use* CHDER *or* CHINT *respectively. Finally, to convert Chebyshev coefficients into coefficients of a polynomial for the same function (a dangerous procedure about which we offer due warning in the text) use* CHEBPC *and* PCSHFT *in succession.*

Chapter 5 also treats several methods for which we supply no programs. These are continued fractions, rational functions, recurrence relations, and the solution of quadratic and cubic equations.

⋆ ⋆ ⋆ ⋆

Subroutine EULSUM applies Euler's transformation to the summation of an alternating series. It is called successively for each term to be summed. Our sample program D5R1 evaluates the approximation

$$\ln(1+x) = x - \frac{x^2}{2} + \frac{x^3}{3} - \frac{x^4}{4} + \cdots \qquad -1 < x < 1$$

It asks how many terms MVAL are to be included in the approximation and then makes MVAL calls to EULSUM. Each time, index j increases and TERM takes the value $(-1)^{j+1}x^j/j$. Both this approximation and the function $\ln(1+x)$ itself are evaluated across the region -1 to 1 for comparison. If MVAL is set less than 1 or more than 40, the program terminates.

```
      PROGRAM D5R1
C     Driver for routine EULSUM
      PARAMETER (NVAL=40)
      DIMENSION WKSP(NVAL)
C     Evaluate ln(1+x)=x-x^2/2+x^3/3-x^4/4...  for -1<x<1
10    WRITE(*,*) 'How many terms in polynomial?'
      WRITE(*,'(1X,A,I2,A)') 'Enter n between 1 and ',NVAL,
     *                '. Enter n=0 to end.'
      READ(*,*) MVAL
      IF ((MVAL.LE.0).OR.(MVAL.GT.NVAL)) STOP
      WRITE(*,'(1X,T9,A1,T18,A6,T28,A10)') 'X','Actual','Polynomial'
      DO 12 I=-8,8,1
        X=I/10.0
        SUM=0.0
        XPOWER=-1
        DO 11 J=1,MVAL
          XPOWER=-X*XPOWER
          TERM=XPOWER/J
          CALL EULSUM(SUM,TERM,J,WKSP)
11      CONTINUE
        WRITE(*,'(3F12.6)') X,LOG(1.0+X),SUM
12    CONTINUE
      GOTO 10
      END
```

DDPOLY evaluates a polynomial and its derivatives, given the coefficients of the polynomial in the form of an input vector. Sample program D5R2 illustrates this for the polynomial:

$$(x-1)^5 = -1 + 5x - 10x^2 + 10x^3 - 5x^4 + x^5$$

(This is a foolish example, of course. No one would knowingly evaluate $(x-1)^5$ by multiplying it out and evaluating terms individually—but it gives us a convenient way to check the result!). Since this is a fifth order polynomial, we set NC, the number of coeffients, to 6, and initialize the array C of coefficients, with C(1) being the constant coefficient and C(6) the highest-order coefficient. There are two loops, one of which evaluates for x values from 0.0 to 2.0, and the other of which stores the value of the function and NC-1 derivatives. D(J,I) keeps the entire array of values for printing. In the second part of the program, the polynomial evaluations are compared with

$$f^{(n-1)}(x) = \frac{5!}{(6-n)!}(x-1.0)^{6-n} \qquad n = 1,\ldots,5$$

```
      PROGRAM D5R2
C     Driver for routine DDPOLY
C     Polynomial (X-1)**5
      PARAMETER(NC=6,NCM1=5,NP=20)
      DIMENSION C(NC),PD(NCM1),D(NCM1,NP)
      CHARACTER A(NCM1)*15
      DATA A/'polynomial:','first deriv:','second deriv:',
     *       'third deriv:','fourth deriv:'/
      DATA C/-1.0,5.0,-10.0,10.0,-5.0,1.0/
      DO 12 I=1,NP
        X=0.1*I
```

```
          CALL DDPOLY(C,NC,X,PD,NC-1)
          DO 11 J=1,NC-1
              D(J,I)=PD(J)
11        CONTINUE
12    CONTINUE
      DO 14 I=1,NC-1
          WRITE(*,'(1X,T7,A)') A(I)
          WRITE(*,'(1X,T13,A,T25,A,T40,A)') 'X','DDPOLY','actual'
          DO 13 J=1,NP
              X=0.1*J
              WRITE(*,'(1X,3F15.6)') X,D(I,J),
     *            FACTRL(NC-1)/FACTRL(NC-I)*((X-1.0)**(NC-I))
13        CONTINUE
          WRITE(*,*) 'press ENTER to continue...'
          READ(*,*)
14    CONTINUE
      END
```

POLDIV divides polynomials. Given the coefficients of a numerator and denominator polynomial, POLDIV returns the coefficients of a quotient and a remainder polynomial. Sample program D5R3 takes

$$\text{Numerator} = \mathtt{U} = -1 + 5x - 10x^2 + 10x^3 - 5x^4 + x^5 = (x-1)^5$$

$$\text{Denominator} = \mathtt{V} = 1 + 3x + 3x^2 + x^3 = (x+1)^3$$

for which we expect

$$\text{Quotient} = \mathtt{Q} = 31 - 8x + x^2$$

$$\text{Remainder} = \mathtt{R} = -32 - 80x - 80x^2$$

The program compares these with the output of POLDIV.

```
      PROGRAM D5R3
C     Driver for routine POLDIV
C     (X-1)**5/(X+1)**3
      PARAMETER(N=6,NV=4)
      DIMENSION U(N),V(NV),Q(N),R(N)
      DATA U/-1.0,5.0,-10.0,10.0,-5.0,1.0/
      DATA V/1.0,3.0,3.0,1.0/
      CALL POLDIV(U,N,V,NV,Q,R)
      WRITE(*,'(//1X,6(7X,A)/)')'X^0','X^1','X^2','X^3','X^4','X^5'
      WRITE(*,*) 'Quotient polynomial coefficients:'
      WRITE(*,'(1X,6F10.2/)') (Q(I),I=1,6)
      WRITE(*,*) 'Expected quotient coefficients:'
      WRITE(*,'(1X,6F10.2///)') 31.0,-8.0,1.0,0.0,0.0,0.0
      WRITE(*,*) 'Remainder polynomial coefficients:'
      WRITE(*,'(1X,4F10.2/)') (R(I),I=1,4)
      WRITE(*,*) 'Expected remainder coefficients:'
      WRITE(*,'(1X,4F10.2//)') -32.0,-80.0,-80.0,0.0
      END
```

The remaining six programs all deal with Chebyshev polynomials. CHEBFT evaluates the coefficients for a Chebyshev polynomial approximation of a function on a specified interval and for a maximum degree N of polynomial. Demonstration program D5R4 uses the function FUNC $= x^2(x^2-2)\sin x$ on the interval $(-\pi/2, \pi/2)$ with the maximum degree of NVAL=40. Notice that CHEBFT is called with this maximum degree specified, even though subsequent evaluations may truncate the Chebyshev series at much lower terms. After we choose the number MVAL of terms in the evaluation, the Chebyshev polynomial is evaluated term by term, for x values between -0.8π and 0.8π, and the result F is compared to the actual function value.

```
      PROGRAM D5R4
C     Driver for routine CHEBFT
      PARAMETER(NVAL=40, PIO2=1.5707963, EPS=1E-6)
      EXTERNAL FUNC
      DIMENSION C(NVAL)
      A=-PIO2
      B=PIO2
      CALL CHEBFT(A,B,C,NVAL,FUNC)
C     Test result
10    WRITE(*,*) 'How many terms in Chebyshev evaluation?'
      WRITE(*,'(1X,A,I2,A)') 'Enter n between 6 and ',NVAL,
     *                 '. Enter n=0 to end.'
      READ(*,*) MVAL
      IF ((MVAL.LE.0).OR.(MVAL.GT.NVAL)) GOTO 20
      WRITE(*,'(1X,T10,A,T19,A,T28,A)') 'X','Actual','Chebyshev fit'
      DO 12 I=-8,8,1
        X=I*PIO2/10.0
        Y=(X-0.5*(B+A))/(0.5*(B-A))
C     Evaluate Chebyshev polynomial without using routine CHEBEV
        T0=1.0
        T1=Y
        F=C(2)*T1+C(1)*0.5
        DO 11 J=3,MVAL
          DUM=T1
          T1=2.0*Y*T1-T0
          T0=DUM
          TERM=C(J)*T1
          F=F+TERM
11      CONTINUE
        WRITE(*,'(1X,3F12.6)') X,FUNC(X),F
12    CONTINUE
      GOTO 10
20    END
      FUNCTION FUNC(X)
      FUNC=(X**2)*(X**2-2.0)*SIN(X)
      END
```

CHEBEV is the Chebyshev polynomial evaluator and the next sample program D5R5 uses it for the same problem just discussed. In fact, the program is identical except that it replaces the internal polynomial summation with CHEBEV, which applies Clenshaw's recurrence to find the polynomial values.

```
      PROGRAM D5R5
C     Driver for routine CHEBEV
      PARAMETER(NVAL=40, PIO2=1.5707963)
      EXTERNAL FUNC
      DIMENSION C(NVAL)
      A=-PIO2
      B=PIO2
      CALL CHEBFT(A,B,C,NVAL,FUNC)
C     Test Chebyshev evaluation routine
10    WRITE(*,*) 'How many terms in Chebyshev evaluation?'
```

```
      WRITE(*,'(1X,A,I2,A)') 'Enter n between 6 and ',NVAL,
     *                     '. Enter n=0 to end.'
      READ(*,*) MVAL
      IF ((MVAL.LE.0).OR.(MVAL.GT.NVAL)) GOTO 20
      WRITE(*,'(1X,T10,A,T19,A,T28,A)') 'X','Actual','Chebyshev fit'
      DO 11 I=-8,8,1
          X=I*PIO2/10.0
          WRITE(*,'(1X,3F12.6)') X,FUNC(X),CHEBEV(A,B,C,MVAL,X)
11    CONTINUE
      GOTO 10
20    END
      FUNCTION FUNC(X)
      FUNC=(X**2)*(X**2-2.0)*SIN(X)
      END
```

By the same token, the tests for CHINT and CHDER needn't be much different. CHINT determines Chebyshev coefficients for the integral of the function, and CHDER for the derivative of the function, given the Chebyshev coefficients for the function itself (from CHEBFT) and the interval (A, B) of evaluation. When applied to the function above, the true integral is

$$\text{FINT} = 4x(x^2 - 7)\sin x - (x^4 - 14x^2 + 28)\cos x$$

and the true derivative is

$$\text{FDER} = 4x(x^2 - 1)\sin x + x^2(x^2 - 2)\cos x$$

The code in sample programs D5R6 and D5R7 compares the true and Chebyshev-derived integral and derivative values for a range of x in the interval of evaluation. Since CHINT and CHDER return Chebyshev coefficients, and not the integral and derivative values themselves, calls to CHEBEV are required for the comparison.

```
      PROGRAM D5R6
C     Driver for routine CHINT
      PARAMETER(NVAL=40, PIO2=1.5707963)
      EXTERNAL FUNC,FINT
      DIMENSION C(NVAL),CINT(NVAL)
      A=-PIO2
      B=PIO2
      CALL CHEBFT(A,B,C,NVAL,FUNC)
C     Test integral
10    WRITE(*,*) 'How many terms in Chebyshev evaluation?'
      WRITE(*,'(1X,A,I2,A)') 'Enter n between 6 and ',NVAL,
     *                     '. Enter n=0 to end.'
      READ(*,*) MVAL
      IF ((MVAL.LE.0).OR.(MVAL.GT.NVAL)) GOTO 20
      CALL CHINT(A,B,C,CINT,MVAL)
      WRITE(*,'(1X,T10,A,T19,A,T29,A)') 'X','Actual','Cheby. Integ.'
      DO 11 I=-8,8,1
          X=I*PIO2/10.0
          WRITE(*,'(1X,3F12.6)') X,FINT(X)-FINT(-PIO2)
     *                    ,CHEBEV(A,B,CINT,MVAL,X)
11    CONTINUE
      GOTO 10
20    END
      FUNCTION FUNC(X)
```

```
      FUNC=(X**2)*(X**2-2.0)*SIN(X)
      END
      FUNCTION FINT(X)
C     Integral of FUNC
      FINT=4.0*X*((X**2)-7.0)*SIN(X)
     *       -((X**4)-14.0*(X**2)+28.0)*COS(X)
      END

      PROGRAM D5R7
C     Driver for routine CHDER
      PARAMETER(NVAL=40, PIO2=1.5707963)
      EXTERNAL FUNC,FDER
      DIMENSION C(NVAL),CDER(NVAL)
      A=-PIO2
      B=PIO2
      CALL CHEBFT(A,B,C,NVAL,FUNC)
C     Test derivative
10    WRITE(*,*) 'How many terms in Chebyshev evaluation?'
      WRITE(*,'(1X,A,I2,A)') 'Enter n between 6 and ',NVAL,
     *            '. Enter n=0 to end.'
      READ(*,*) MVAL
      IF ((MVAL.LE.0).OR.(MVAL.GT.NVAL)) GOTO 20
      CALL CHDER(A,B,C,CDER,MVAL)
      WRITE(*,'(1X,T10,A,T19,A,T28,A)') 'X','Actual','Cheby. Deriv.'
      DO 11 I=-8,8,1
        X=I*PIO2/10.0
        WRITE(*,'(1X,3F12.6)') X,FDER(X),CHEBEV(A,B,CDER,MVAL,X)
11    CONTINUE
      GOTO 10
20    END
      FUNCTION FUNC(X)
      FUNC=(X**2)*(X**2-2.0)*SIN(X)
      END
      FUNCTION FDER(X)
C     Derivative of FUNC
      FDER=4.0*X*((X**2)-1.0)*SIN(X)+(X**2)*(X**2-2.0)*COS(X)
      END
```

The final two programs of this chapter turn the coefficients of a Chebyshev approximation into those of a polynomial approximation in the variable

$$y = \frac{x - \frac{1}{2}(B+A)}{\frac{1}{2}(B-A)}$$

(routine CHEBPC), or of a polynomial approximation in x itself (routine CHEBPC followed by PCSHFT). These procedures are discouraged for reasons discussed in *Numerical Recipes*, but should they serve some special purpose for you, we have at least warned that you will be sacrificing accuracy, particularly for polynomials above order 7 or 8. Sample program D5R8 calls CHEBFT and CHEBPC to find polynomial coefficients in y for a truncated series. For a set of x values between $-\pi$ and π it calculates y and then the terms of the y-polynomial, which are summed in variable POLY. Finally, POLY is compared to the true function value. (The function FUNC is the same used before.)

```
      PROGRAM D5R8
C     Driver for routine CHEBPC
      PARAMETER(NVAL=40, PIO2=1.5707963)
      EXTERNAL FUNC
      DIMENSION C(NVAL),D(NVAL)
      A=-PIO2
      B=PIO2
      CALL CHEBFT(A,B,C,NVAL,FUNC)
10    WRITE(*,*) 'How many terms in Chebyshev evaluation?'
      WRITE(*,'(1X,A,I2,A)') 'Enter n between 6 and ',NVAL,
     *                '. Enter n=0 to end.'
      READ(*,*) MVAL
      IF ((MVAL.LE.0).OR.(MVAL.GT.NVAL)) GOTO 20
      CALL CHEBPC(C,D,MVAL)
C     Test polynomial
      WRITE(*,'(1X,T10,A,T19,A,T29,A)') 'X','Actual','Polynomial'
      DO 12 I=-8,8,1
         X=I*PIO2/10.0
         Y=(X-(0.5*(B+A)))/(0.5*(B-A))
         POLY=D(MVAL)
         DO 11 J=MVAL-1,1,-1
            POLY=POLY*Y+D(J)
11       CONTINUE
         WRITE(*,'(1X,3F12.6)') X,FUNC(X),POLY
12    CONTINUE
      GOTO 10
20    END
      FUNCTION FUNC(X)
      FUNC=(X**2)*(X**2-2.0)*SIN(X)
      END
```

PCSHFT shifts the polynomial to be one in variable x. Sample program D5R9 is like the previous program except that it follows the call to CHEBPC with a call to PCSHFT.

```
      PROGRAM D5R9
C     Driver for routine PCSHFT
      PARAMETER(NVAL=40, PIO2=1.5707963)
      EXTERNAL FUNC
      DIMENSION C(NVAL),D(NVAL)
      A=-PIO2
      B=PIO2
      CALL CHEBFT(A,B,C,NVAL,FUNC)
10    WRITE(*,*) 'How many terms in Chebyshev evaluation?'
      WRITE(*,'(1X,A,I2,A)') 'Enter n between 6 and ',NVAL,
     *                '. Enter n=0 to end.'
      READ(*,*) MVAL
      IF ((MVAL.LE.0).OR.(MVAL.GT.NVAL)) GOTO 20
      CALL CHEBPC(C,D,MVAL)
      CALL PCSHFT(A,B,D,MVAL)
C     Test shifted polynomial
      WRITE(*,'(1X,T10,A,T19,A,T29,A)') 'X','Actual','Polynomial'
      DO 12 I=-8,8,1
         X=I*PIO2/10.0
         POLY=D(MVAL)
         DO 11 J=MVAL-1,1,-1
            POLY=POLY*X+D(J)
```

```
11          CONTINUE
            WRITE(*,'(1X,3F12.6)') X,FUNC(X),POLY
12        CONTINUE
          GOTO 10
20        END
          FUNCTION FUNC(X)
          FUNC=(X**2)*(X**2-2.0)*SIN(X)
          END
```

Chapter 6: Special Functions

This chapter on special functions provides illustrations of techniques developed in Chapter 5. At the same time, it offers routines for calculating many of the functions that arise frequently in analytical work, but which are not so common to be included, for example, as a single keystroke on your pocket calculator. In terms of demonstration programs, they represent a simple bunch. The test routines are all virtually identical, all making reference to a single file of function values called `FNCVAL.DAT` *which is listed in the Appendix at the end of this chapter. In this file are accurate values for the individual functions for a variety of values for each argument. We have aimed to "stress" the routines a bit by throwing in some extreme values for the arguments.*

Many of the function values came from Abramowitz and Stegun's Handbook of Mathematical Functions. Some others, however, came from our library of dusty volumes from past masters. There is an implicit danger in a comparison test like this—namely, that our source has used the same algorithms as ours to construct the tables. In that case, we test only our mutual competence at computing, not the correctness of the result. Nevertheless, there is some assurance in knowing that the values we calculate are the ones that have been used and scrutinized for many years. Moreover, the expressions for the functions themselves can be worked out in certain special or limiting cases without computer aid, and in these instances the results have proven correct.

⋆ ⋆ ⋆ ⋆

With few exceptions, the routines that follow work in this fashion:

1. Open file `FNCVAL.DAT`.
2. Find the appropriate data table according to its title.
3. Read the argument list for each table entry and pass them to the routine to be tested.
4. Print the arguments along with the expected and actual results.

For the routines in this list, therefore, we forego any further comment, but simply identify them by the special function that they evaluate.

Natural logarithm of the gamma function for positive arguments:

```
      PROGRAM D6R1
C     Driver for routine GAMMLN
      CHARACTER TEXT*14
```

```
      PARAMETER(PI=3.1415926)
      OPEN(5,FILE='FNCVAL.DAT',STATUS='OLD')
10    READ(5,'(A)') TEXT
      IF (TEXT.NE.'Gamma Function') GOTO 10
      READ(5,*) NVAL
      WRITE(*,*) 'Log of gamma function:'
      WRITE(*,'(1X,T11,A1,T24,A6,T40,A10)')
     *        'X','Actual','GAMMLN(X)'
      DO 11 I=1,NVAL
        READ(5,*) X,ACTUAL
        IF (X.GT.0.0) THEN
          IF (X.GE.1.0) THEN
            CALC=GAMMLN(X)
          ELSE
            CALC=GAMMLN(X+1.0)-LOG(X)
          ENDIF
          WRITE(*,'(F12.2,2F18.6)') X,LOG(ACTUAL),CALC
        ENDIF
11    CONTINUE
      CLOSE(5)
      END
```

Factorial function $N!$:

```
      PROGRAM D6R2
C     Driver for routine FACTRL
      CHARACTER TEXT*11
      OPEN(5,FILE='FNCVAL.DAT',STATUS='OLD')
10    READ(5,'(A)') TEXT
      IF (TEXT.NE.'N-factorial') GOTO 10
      READ(5,*) NVAL
      WRITE(*,*) TEXT
      WRITE(*,'(1X,T6,A1,T21,A6,T38,A9)')
     *        'N','Actual','FACTRL(N)'
      DO 11 I=1,NVAL
        READ(5,*) N,ACTUAL
        IF (ACTUAL.LT.(1.0E10)) THEN
          WRITE(*,'(I6,2F20.0)') N,ACTUAL,FACTRL(N)
        ELSE
          WRITE(*,'(I6,2E20.7)') N,ACTUAL,FACTRL(N)
        ENDIF
11    CONTINUE
      CLOSE(5)
      END
```

Binomial coefficients:

```
      PROGRAM D6R3
C     Driver for routine BICO
      CHARACTER TEXT*21
      OPEN(5,FILE='FNCVAL.DAT',STATUS='OLD')
10    READ(5,'(A)') TEXT
      IF (TEXT.NE.'Binomial Coefficients') GOTO 10
      READ(5,*) NVAL
      WRITE(*,*) TEXT
      WRITE(*,'(1X,T6,A1,T12,A1,T19,A6,T28,A9)')
     *        'N','K','Actual','BICO(N,K)'
      DO 11 I=1,NVAL
```

```
            READ(5,*) N,K,BINCO
            WRITE(*,'(2I6,2F12.0)') N,K,BINCO,BICO(N,K)
11      CONTINUE
        CLOSE(5)
        END
```

Natural logarithm of $N!$:

```
        PROGRAM D6R4
C       Driver for routine FACTLN
        CHARACTER TEXT*11
        OPEN(5,FILE='FNCVAL.DAT',STATUS='OLD')
10      READ(5,'(A)') TEXT
        IF (TEXT.NE.'N-factorial') GOTO 10
        READ(5,*) NVAL
        WRITE(*,*) 'Log of N-factorial'
        WRITE(*,'(1X,T6,A1,T18,A6,T34,A9)')
     *          'N','Actual','FACTLN(N)'
        DO 11 I=1,NVAL
            READ(5,*) N,VALUE
            WRITE(*,'(I6,2F18.6)') N,LOG(VALUE),FACTLN(N)
11      CONTINUE
        CLOSE(5)
        END
```

Beta function:

```
        PROGRAM D6R5
C       Driver for routine BETA
        CHARACTER TEXT*13
        OPEN(5,FILE='FNCVAL.DAT',STATUS='OLD')
10      READ(5,'(A)') TEXT
        IF (TEXT.NE.'Beta Function') GOTO 10
        READ(5,*) NVAL
        WRITE(*,*) TEXT
        WRITE(*,'(1X,T5,A1,T11,A1,T24,A6,T43,A9)')
     *          'W','Z','Actual','BETA(W,Z)'
        DO 11 I=1,NVAL
            READ(5,*) W,Z,VALUE
            WRITE(*,'(2F6.2,2E20.6)') W,Z,VALUE,BETA(W,Z)
11      CONTINUE
        CLOSE(5)
        END
```

Incomplete gamma function $P(a,x)$:

```
        PROGRAM D6R6
C       Driver for routine GAMMP
        CHARACTER TEXT*25
        OPEN(5,FILE='FNCVAL.DAT',STATUS='OLD')
10      READ(5,'(A)') TEXT
        IF (TEXT.NE.'Incomplete Gamma Function') GOTO 10
        READ(5,*) NVAL
        WRITE(*,*) TEXT
        WRITE(*,'(1X,T5,A,T16,A,T25,A,T35,A)')
     *          'A','X','Actual','GAMMP(A,X)'
        DO 11 I=1,NVAL
            READ(5,*) A,X,VALUE
```

```
          WRITE(*,'(1X,F6.2,3F12.6)') A,X,VALUE,GAMMP(A,X)
11      CONTINUE
        CLOSE(5)
        END
```

Incomplete gamma function $Q(a, x) = 1 - P(a, x)$:

```
        PROGRAM D6R7
C       Driver for routine GAMMQ
        CHARACTER TEXT*25
        OPEN(5,FILE='FNCVAL.DAT',STATUS='OLD')
10      READ(5,'(A)') TEXT
        IF (TEXT.NE.'Incomplete Gamma Function') GOTO 10
        READ(5,*) NVAL
        WRITE(*,*) TEXT
        WRITE(*,'(1X,T5,A,T16,A,T25,A,T35,A)')
     *          'A','X','Actual','GAMMQ(A,X)'
        DO 11 I=1,NVAL
          READ(5,*) A,X,VALUE
          WRITE(*,'(1X,F6.2,3F12.6)') A,X,1.0-VALUE,GAMMQ(A,X)
11      CONTINUE
        CLOSE(5)
        END
```

Incomplete gamma function $P(a, x)$ evaluated from series representation:

```
        PROGRAM D6R8
C       Driver for routine GSER
        CHARACTER TEXT*25
        OPEN(5,FILE='FNCVAL.DAT',STATUS='OLD')
10      READ(5,'(A)') TEXT
        IF (TEXT.NE.'Incomplete Gamma Function') GOTO 10
        READ(5,*) NVAL
        WRITE(*,*) TEXT
        WRITE(*,'(1X,T5,A,T16,A,T25,A,T36,A,T47,A,T62,A)')
     *          'A','X','Actual','GSER(A,X)','GAMMLN(A)','GLN'
        DO 11 I=1,NVAL
          READ(5,*) A,X,VALUE
          CALL GSER(GAMSER,A,X,GLN)
          WRITE(*,'(1X,F6.2,5F12.6)') A,X,VALUE,GAMSER,
     *          GAMMLN(A),GLN
11      CONTINUE
        CLOSE(5)
        END
```

Incomplete gamma function $Q(a, x)$ evaluated by continued fraction representation:

```
        PROGRAM D6R9
C       Driver for routine GCF
        CHARACTER TEXT*25
        OPEN(5,FILE='FNCVAL.DAT',STATUS='OLD')
10      READ(5,'(A)') TEXT
        IF (TEXT.NE.'Incomplete Gamma Function') GOTO 10
        READ(5,*) NVAL
        WRITE(*,*) TEXT
        WRITE(*,'(1X,T5,A,T16,A,T25,A,T36,A,T47,A,T62,A)')
     *          'A','X','Actual','GCF(A,X)','GAMMLN(A)','GLN'
        DO 11 I=1,NVAL
```

```
          READ(5,*) A,X,VALUE
          IF (X.GE.A+1.0) THEN
            CALL GCF(GAMMCF,A,X,GLN)
            WRITE(*,'(1X,F6.2,5F12.6)') A,X,1.0-VALUE,
     *                GAMMCF,GAMMLN(A),GLN
          ENDIF
11      CONTINUE
        CLOSE(5)
        END
```

Error function:

```
      PROGRAM D6R10
C     Driver for routine ERF
      CHARACTER TEXT*14
      OPEN(5,FILE='FNCVAL.DAT',STATUS='OLD')
10    READ(5,'(A)') TEXT
      IF (TEXT.NE.'Error Function') GOTO 10
      READ(5,*) NVAL
      WRITE(*,*) TEXT
      WRITE(*,'(1X,T5,A1,T12,A6,T24,A6)')
     *      'X','Actual','ERF(X)'
      DO 11 I=1,NVAL
        READ(5,*) X,VALUE
        WRITE(*,'(F6.2,2F12.7)') X,VALUE,ERF(X)
11    CONTINUE
      CLOSE(5)
      END
```

Complementary error function:

```
      PROGRAM D6R11
C     Driver for routine ERFC
      CHARACTER TEXT*14
      OPEN(5,FILE='FNCVAL.DAT',STATUS='OLD')
10    READ(5,'(A)') TEXT
      IF (TEXT.NE.'Error Function') GOTO 10
      READ(5,*) NVAL
      WRITE(*,*) 'Complementary error function'
      WRITE(*,'(1X,T5,A1,T12,A6,T23,A7)')
     *      'X','Actual','ERFC(X)'
      DO 11 I=1,NVAL
        READ(5,*) X,VALUE
        VALUE=1.0-VALUE
        WRITE(*,'(F6.2,2F12.7)') X,VALUE,ERFC(X)
11    CONTINUE
      CLOSE(5)
      END
```

Complementary error function from a Chebyshev fit to a guessed functional form:

```
      PROGRAM D6R12
C     Driver for routine ERFCC
      CHARACTER TEXT*14
      OPEN(5,FILE='FNCVAL.DAT',STATUS='OLD')
10    READ(5,'(A)') TEXT
      IF (TEXT.NE.'Error Function') GOTO 10
      READ(5,*) NVAL
```

```
      WRITE(*,*) 'Complementary error function'
      WRITE(*,'(1X,T5,A1,T12,A6,T23,A8)')
     *         'X','Actual','ERFCC(X)'
      DO 11 I=1,NVAL
          READ(5,*) X,VALUE
          VALUE=1.0-VALUE
          WRITE(*,'(F6.2,2F12.7)') X,VALUE,ERFCC(X)
11    CONTINUE
      CLOSE(5)
      END
```

Incomplete Beta function:

```
      PROGRAM D6R13
C     Driver for routine BETAI,BETACF
      CHARACTER TEXT*24
      OPEN(5,FILE='FNCVAL.DAT',STATUS='OLD')
10    READ(5,'(A)') TEXT
      IF (TEXT.NE.'Incomplete Beta Function') GOTO 10
      READ(5,*) NVAL
      WRITE(*,*) TEXT
      WRITE(*,'(1X,T5,A,T15,A,T27,A,T36,A,T47,A)')
     *         'A','B','X','Actual','BETAI(X)'
      DO 11 I=1,NVAL
          READ(5,*) A,B,X,VALUE
          WRITE(*,'(F6.2,4F12.6)') A,B,X,VALUE,BETAI(A,B,X)
11    CONTINUE
      CLOSE(5)
      END
```

Bessel function J_0:

```
      PROGRAM D6R15
C     Driver for routine BESSJ0
      CHARACTER TEXT*18
      OPEN(5,FILE='FNCVAL.DAT',STATUS='OLD')
10    READ(5,'(A)') TEXT
      IF (TEXT.NE.'Bessel Function J0') GOTO 10
      READ(5,*) NVAL
      WRITE(*,*) TEXT
      WRITE(*,'(1X,T5,A1,T12,A6,T22,A9)')
     *         'X','Actual','BESSJ0(X)'
      DO 11 I=1,NVAL
          READ(5,*) X,VALUE
          WRITE(*,'(F6.2,2F12.7)') X,VALUE,BESSJ0(X)
11    CONTINUE
      CLOSE(5)
      END
```

Bessel function Y_0:

```
      PROGRAM D6R16
C     Driver for routine BESSY0
      CHARACTER TEXT*18
      OPEN(5,FILE='FNCVAL.DAT',STATUS='OLD')
10    READ(5,'(A)') TEXT
      IF (TEXT.NE.'Bessel Function Y0') GOTO 10
      READ(5,*) NVAL
```

```
      WRITE(*,*) TEXT
      WRITE(*,'(1X,T5,A1,T12,A6,T22,A9)')
     *         'X','Actual','BESSY0(X)'
      DO 11 I=1,NVAL
        READ(5,*) X,VALUE
        WRITE(*,'(F6.2,2F12.7)') X,VALUE,BESSY0(X)
11    CONTINUE
      CLOSE(5)
      END
```

Bessel function J_1:

```
      PROGRAM D6R17
C     Driver for routine BESSJ1
      CHARACTER TEXT*18
      OPEN(5,FILE='FNCVAL.DAT',STATUS='OLD')
10    READ(5,'(A)') TEXT
      IF (TEXT.NE.'Bessel Function J1') GOTO 10
      READ(5,*) NVAL
      WRITE(*,*) TEXT
      WRITE(*,'(1X,T5,A1,T12,A6,T22,A9)')
     *         'X','Actual','BESSJ1(X)'
      DO 11 I=1,NVAL
        READ(5,*) X,VALUE
        WRITE(*,'(F6.2,2F12.7)') X,VALUE,BESSJ1(X)
11    CONTINUE
      CLOSE(5)
      END
```

Bessel function Y_1:

```
      PROGRAM D6R18
C     Driver for routine BESSY1
      CHARACTER TEXT*18
      OPEN(5,FILE='FNCVAL.DAT',STATUS='OLD')
10    READ(5,'(A)') TEXT
      IF (TEXT.NE.'Bessel Function Y1') GOTO 10
      READ(5,*) NVAL
      WRITE(*,*) TEXT
      WRITE(*,'(1X,T5,A1,T12,A6,T22,A9)')
     *         'X','Actual','BESSY1(X)'
      DO 11 I=1,NVAL
        READ(5,*) X,VALUE
        WRITE(*,'(F6.2,2F12.7)') X,VALUE,BESSY1(X)
11    CONTINUE
      CLOSE(5)
      END
```

Bessel function Y_n for $n > 1$:

```
      PROGRAM D6R19
C     Driver for routine BESSY
      CHARACTER TEXT*18
      OPEN(5,FILE='FNCVAL.DAT',STATUS='OLD')
10    READ(5,'(A)') TEXT
      IF (TEXT.NE.'Bessel Function Yn') GOTO 10
      READ(5,*) NVAL
      WRITE(*,*) TEXT
```

```
      WRITE(*,'(1X,T5,A,T12,A,T20,A,T33,A)')
     *         'N','X','Actual','BESSY(N,X)'
      DO 11 I=1,NVAL
          READ(5,*) N,X,VALUE
          WRITE(*,'(1X,I4,F8.2,2E15.6)') N,X,VALUE,BESSY(N,X)
11    CONTINUE
      CLOSE(5)
      END
```

Bessel function J_n for $n > 1$:

```
      PROGRAM D6R20
C     Driver for routine BESSJ
      CHARACTER TEXT*18
      OPEN(5,FILE='FNCVAL.DAT',STATUS='OLD')
10    READ(5,'(A)') TEXT
      IF (TEXT.NE.'Bessel Function Jn') GOTO 10
      READ(5,*) NVAL
      WRITE(*,*) TEXT
      WRITE(*,'(1X,T5,A,T12,A,T20,A,T33,A)')
     *         'N','X','Actual','BESSJ(N,X)'
      DO 11 I=1,NVAL
          READ(5,*) N,X,VALUE
          WRITE(*,'(1X,I4,F8.2,2E15.6)') N,X,VALUE,BESSJ(N,X)
11    CONTINUE
      CLOSE(5)
      END
```

Bessel function I_0:

```
      PROGRAM D6R21
C     Driver for routine BESSI0
      CHARACTER TEXT*27
      OPEN(5,FILE='FNCVAL.DAT',STATUS='OLD')
10    READ(5,'(A)') TEXT
      IF (TEXT.NE.'Modified Bessel Function I0') GOTO 10
      READ(5,*) NVAL
      WRITE(*,*) TEXT
      WRITE(*,'(1X,T5,A,T13,A,T28,A)')
     *         'X','Actual','BESSI0(X)'
      DO 11 I=1,NVAL
          READ(5,*) X,VALUE
          WRITE(*,'(F6.2,2E16.7)') X,VALUE,BESSI0(X)
11    CONTINUE
      CLOSE(5)
      END
```

Bessel function K_0:

```
      PROGRAM D6R22
C     Driver for routine BESSK0
      CHARACTER TEXT*27
      OPEN(5,FILE='FNCVAL.DAT',STATUS='OLD')
10    READ(5,'(A)') TEXT
      IF (TEXT.NE.'Modified Bessel Function K0') GOTO 10
      READ(5,*) NVAL
      WRITE(*,*) TEXT
      WRITE(*,'(1X,T5,A,T13,A,T28,A)')
```

```
     *         'X','Actual','BESSK0(X)'
      DO 11 I=1,NVAL
          READ(5,*) X,VALUE
          WRITE(*,'(F6.2,2E16.7)') X,VALUE,BESSK0(X)
11    CONTINUE
      CLOSE(5)
      END
```

Bessel function I_1:

```
      PROGRAM D6R23
C     Driver for routine BESSI1
      CHARACTER TEXT*27
      OPEN(5,FILE='FNCVAL.DAT',STATUS='OLD')
10    READ(5,'(A)') TEXT
      IF (TEXT.NE.'Modified Bessel Function I1') GOTO 10
      READ(5,*) NVAL
      WRITE(*,*) TEXT
      WRITE(*,'(1X,T5,A,T13,A,T28,A)')
     *         'X','Actual','BESSI1(X)'
      DO 11 I=1,NVAL
          READ(5,*) X,VALUE
          WRITE(*,'(F6.2,2E16.7)') X,VALUE,BESSI1(X)
11    CONTINUE
      CLOSE(5)
      END
```

Bessel function K_1:

```
      PROGRAM D6R24
C     Driver for routine BESSK1
      CHARACTER TEXT*27
      OPEN(5,FILE='FNCVAL.DAT',STATUS='OLD')
10    READ(5,'(A)') TEXT
      IF (TEXT.NE.'Modified Bessel Function K1') GOTO 10
      READ(5,*) NVAL
      WRITE(*,*) TEXT
      WRITE(*,'(1X,T5,A,T13,A,T28,A)')
     *         'X','Actual','BESSK1(X)'
      DO 11 I=1,NVAL
          READ(5,*) X,VALUE
          WRITE(*,'(F6.2,2E16.7)') X,VALUE,BESSK1(X)
11    CONTINUE
      CLOSE(5)
      END
```

Bessel function K_n for $n > 1$:

```
      PROGRAM D6R25
C     Driver for routine BESSK
      CHARACTER TEXT*27
      OPEN(5,FILE='FNCVAL.DAT',STATUS='OLD')
10    READ(5,'(A)') TEXT
      IF (TEXT.NE.'Modified Bessel Function Kn') GOTO 10
      READ(5,*) NVAL
      WRITE(*,*) TEXT
      WRITE(*,'(1X,T5,A,T12,A,T20,A,T35,A)')
     *         'N','X','Actual','BESSK(N,X)'
```

```
      DO 11 I=1,NVAL
          READ(5,*) N,X,VALUE
          WRITE(*,'(1X,I4,F8.2,2E16.7)') N,X,VALUE,BESSK(N,X)
11    CONTINUE
      CLOSE(5)
      END
```

Bessel function I_n for $n > 1$:

```
      PROGRAM D6R26
C     Driver for routine BESSI
      CHARACTER TEXT*27
      OPEN(5,FILE='FNCVAL.DAT',STATUS='OLD')
10    READ(5,'(A)') TEXT
      IF (TEXT.NE.'Modified Bessel Function In') GOTO 10
      READ(5,*) NVAL
      WRITE(*,*) TEXT
      WRITE(*,'(1X,T5,A,T12,A,T20,A,T34,A)')
     *        'N','X','Actual','BESSI(N,X)'
      DO 11 I=1,NVAL
          READ(5,*) N,X,VALUE
          WRITE(*,'(1X,I4,F8.2,2E16.7)') N,X,VALUE,BESSI(N,X)
11    CONTINUE
      CLOSE(5)
      END
```

Legendre polynomials:

```
      PROGRAM D6R27
C     Driver for routine PLGNDR
      CHARACTER TEXT*27
      OPEN(5,FILE='FNCVAL.DAT',STATUS='OLD')
10    READ(5,'(A)',ERR=99) TEXT
      IF (TEXT.EQ.'Legendre Polynomials') THEN
          READ(5,*) NVAL
          WRITE(*,*) TEXT
          WRITE(*,'(1X,T5,A,T9,A,T20,A,T35,A,T49,A)')
     *            'N','M','X','Actual','PLGNDR(N,M,X)'
          DO 12 I=1,NVAL
              READ(5,*) N,M,X,VALUE
              FAC=1.0
              IF (M.GT.0) THEN
                  DO 11 J=N-M+1,N+M
                      FAC=FAC*J
11                CONTINUE
              ENDIF
              FAC=2.0*FAC/(2.0*N+1.0)
              VALUE=VALUE*SQRT(FAC)
              WRITE(*,'(1X,2I4,3E17.6)') N,M,X,VALUE,PLGNDR(N,M,X)
12        CONTINUE
      ENDIF
      GOTO 10
99    CLOSE(5)
      END
```

Jacobian elliptic functions:

```
      PROGRAM D6R30
C     Driver for routine SNCNDN
      CHARACTER TEXT*26
      OPEN(5,FILE='FNCVAL.DAT',STATUS='OLD')
10    READ(5,'(A)',ERR=99) TEXT
      IF (TEXT.EQ.'Jacobian Elliptic Function') THEN
          READ(5,*) NVAL
          WRITE(*,*) TEXT
          WRITE(*,'(1X,T4,A,T13,A,T21,A,T38,A,T49,A,T60,A)')
     *            'Mc','U','Actual','SN','SN^2+CN^2',
     *            '(Mc)*(SN^2)+DN^2'
          DO 11 I=1,NVAL
              READ(5,*) EM,UU,VALUE
              EMMC=1.0-EM
              CALL SNCNDN(UU,EMMC,SN,CN,DN)
              RESULT1=SN*SN+CN*CN
              RESULT2=EM*SN*SN+DN*DN
              WRITE(*,'(1X,F5.2,F8.2,2E15.5,F12.5,F14.5)')
     *                EMMC,UU,VALUE,SN,RESULT1,RESULT2
11        CONTINUE
      ENDIF
      GOTO 10
99    CLOSE(5)
      END
```

There are three programs that operate in a slightly different fashion. Sample programs D6R28 and D6R29 for routines EL2 and CEL, which calculate elliptic integrals, do not refer to tables at all. Instead, they make twenty random choices of argument and compare the output of the function evaluation routine for these arguments to the result of actually performing the integration that defines them. The routine QSIMP from Chapter 4 of *Numerical Recipes* is used for the integration.

```
      PROGRAM D6R28
C     Driver for routine EL2
      COMMON AKC,A,B
      EXTERNAL FUNC
      WRITE(*,*) 'General Elliptic Integral of Second Kind'
      WRITE(*,'(1X,T8,A,T17,A,T28,A,T38,A,T47,A,T54,A)')
     *        'x','kc','a','b','EL2','Integral'
      IDUM=-55
      AGO=0.0
      DO 11 I=1,20
          AKC=5.0*RAN3(IDUM)
          A=10.0*RAN3(IDUM)
          B=10.0*RAN3(IDUM)
          X=10.0*RAN3(IDUM)
          ASTOP=ATAN(X)
          CALL QSIMP(FUNC,AGO,ASTOP,S)
          WRITE(*,'(1X,6F10.6)')
     *            X,AKC,A,B,EL2(X,AKC,A,B),S
11    CONTINUE
      END
      FUNCTION FUNC(PHI)
      COMMON AKC,A,B
      TN=TAN(PHI)
      TSQ=TN*TN
```

```
      FUNC=(A+B*TSQ)/SQRT((1.0+TSQ)*(1.0+AKC*AKC*TSQ))
      END

      PROGRAM D6R29
C     Driver for routine CEL
      COMMON A,B,P,AKC
      EXTERNAL FUNC
      PARAMETER(PIO2=1.5707963)
      WRITE(*,*) 'Complete Elliptic Integral'
      WRITE(*,'(1X,T7,A,T18,A,T28,A,T38,A,T47,A,T54,A)')
     *        'kc','p','a','b','CEL','Integral'
      IDUM=-55
      AGO=0.0
      ASTOP=PIO2
      DO 11 I=1,20
        AKC=0.1+RAN3(IDUM)
        A=10.0*RAN3(IDUM)
        B=10.0*RAN3(IDUM)
        P=0.1+RAN3(IDUM)
        CALL QSIMP(FUNC,AGO,ASTOP,S)
        WRITE(*,'(1X,6F10.6)')
     *          AKC,P,A,B,CEL(AKC,P,A,B),S
11    CONTINUE
      END
      FUNCTION FUNC(PHI)
      COMMON A,B,P,AKC
      CS=COS(PHI)
      CSQ=CS*CS
      SSQ=1.0-CSQ
      FUNC=(A*CSQ+B*SSQ)/(CSQ+P*SSQ)/SQRT(CSQ+AKC*AKC*SSQ)
      END
```

Routine SNCNDN returns Jacobian elliptic functions. The file FNCVAL.DAT contains information only about function SN. However, the values of CN and DN satisfy the relationships

$$\mathrm{SN}^2 + \mathrm{CN}^2 = 1, \qquad k^2\mathrm{SN}^2 + \mathrm{DN}^2 = 1.$$

The program D6R30 works exactly as the others in terms of testing SN, but for verifying CN and DN it lists the values RESULT1 and RESULT2 of the left sides of the two equations above. Each of them should have the value 1.0 for all choices of arguments.

```
      PROGRAM D6R30
C     Driver for routine SNCNDN
      CHARACTER TEXT*26
      OPEN(5,FILE='FNCVAL.DAT',STATUS='OLD')
10    READ(5,'(A)',ERR=99) TEXT
      IF (TEXT.EQ.'Jacobian Elliptic Function') THEN
        READ(5,*) NVAL
        WRITE(*,*) TEXT
        WRITE(*,'(1X,T4,A,T13,A,T21,A,T38,A,T49,A,T60,A)')
     *          'Mc','U','Actual','SN','SN^2+CN^2',
     *          '(Mc)*(SN^2)+DN^2'
        DO 11 I=1,NVAL
          READ(5,*) EM,UU,VALUE
          EMMC=1.0-EM
          CALL SNCNDN(UU,EMMC,SN,CN,DN)
```

```
               RESULT1=SN*SN+CN*CN
               RESULT2=EM*SN*SN+DN*DN
               WRITE(*,'(1X,F5.2,F8.2,2E15.5,F12.5,F14.5)')
     *                EMMC,UU,VALUE,SN,RESULT1,RESULT2
11          CONTINUE
        ENDIF
        GOTO 10
99      CLOSE(5)
        END
```

Appendix

File FNCVAL.DAT:

```
Values of Special Functions in format x,F(x) or x,y,F(x,y)
Gamma Function
17 Values
 1.0     1.000000
 1.2     0.918169
 1.4     0.887264
 1.6     0.893515
 1.8     0.931384
 2.0     1.000000
 0.2     4.590845
 0.4     2.218160
 0.6     1.489192
 0.8     1.164230
-0.2     5.2005665E01
-0.4     4.617091E01
-0.6     4.0128959E01
-0.8     3.4231564E01
10.0     3.6288000E05
20.0     1.2164510E17
30.0     8.8417620E30
N-factorial
18 Values
1       1
2       2
3       6
4       24
5       120
6       720
7       5040
8       40320
9       362880
10      3628800
11      39916800
12      479001600
13      6227020800
14      87178291200
15      1.3076755E12
20      2.4329042E18
25      1.5511222E25
30      2.6525281E32
Binomial Coefficients
20 Values
1       0       1
```

```
6       1       6
6       3       20
6       5       6
15      1       15
15      3       455
15      5       3003
15      7       6435
15      9       5005
15      11      1365
15      13      105
25      1       25
25      3       2300
25      5       53130
25      7       480700
25      9       2042975
25      11      4457400
25      13      5200300
25      15      3268760
25      17      1081575
Beta Function
15 Values
1.0     1.0     1.000000
0.2     1.0     5.000000
1.0     0.2     5.000000
0.4     1.0     2.500000
1.0     0.4     2.500000
0.6     1.0     1.666667
0.8     1.0     1.250000
6.0     6.0     3.607504E-04
6.0     5.0     7.936508E-04
6.0     4.0     1.984127E-03
6.0     3.0     5.952381E-03
6.0     2.0     0.238095E-01
7.0     7.0     8.325008E-05
5.0     5.0     1.587302E-03
4.0     4.0     7.142857E-03
3.0     3.0     0.333333E-01
2.0     2.0     1.666667E-01
Incomplete Gamma Function
20 Values
0.1     3.1622777E-02   0.7420263
0.1     3.1622777E-01   0.9119753
0.1     1.5811388       0.9898955
0.5     7.0710678E-02   0.2931279
0.5     7.0710678E-01   0.7656418
0.5     3.5355339       0.9921661
1.0     0.1000000       0.0951626
1.0     1.0000000       0.6321206
1.0     5.0000000       0.9932621
1.1     1.0488088E-01   0.0757471
1.1     1.0488088       0.6076457
1.1     5.2440442       0.9933425
2.0     1.4142136E-01   0.0091054
2.0     1.4142136       0.4130643
2.0     7.0710678       0.9931450
6.0     2.4494897       0.0387318
6.0     12.247449       0.9825937
```

```
11.0   16.583124        0.9404267
26.0   25.495098        0.4863866
41.0   44.821870        0.7359709
Error Function
20 Values
0.0    0.000000
0.1    0.1124629
0.2    0.2227026
0.3    0.3286268
0.4    0.4283924
0.5    0.5204999
0.6    0.6038561
0.7    0.6778012
0.8    0.7421010
0.9    0.7969082
1.0    0.8427008
1.1    0.8802051
1.2    0.9103140
1.3    0.9340079
1.4    0.9522851
1.5    0.9661051
1.6    0.9763484
1.7    0.9837905
1.8    0.9890905
1.9    0.9927904
Incomplete Beta Function
20 Values
0.5    0.5      0.01     0.0637686
0.5    0.5      0.10     0.2048328
0.5    0.5      1.00     1.0000000
1.0    0.5      0.01     0.0050126
1.0    0.5      0.10     0.0513167
1.0    0.5      1.00     1.0000000
1.0    1.0      0.5      0.5000000
5.0    5.0      0.5      0.5000000
10.0   0.5      0.9      0.1516409
10.0   5.0      0.5      0.0897827
10.0   5.0      1.0      1.0000000
10.0   10.0     0.5      0.5000000
20.0   5.0      0.8      0.4598773
20.0   10.0     0.6      0.2146816
20.0   10.0     0.8      0.9507365
20.0   20.0     0.5      0.5000000
20.0   20.0     0.6      0.8979414
30.0   10.0     0.7      0.2241297
30.0   10.0     0.8      0.7586405
40.0   20.0     0.7      0.7001783
Bessel Function J0
20 Values
-5.0   -0.1775968
-4.0   -0.3971498
-3.0   -0.2600520
-2.0    0.2238908
-1.0    0.7651976
 0.0    1.0000000
 1.0    0.7651977
 2.0    0.2238908
```

```
 3.0   -0.2600520
 4.0   -0.3971498
 5.0   -0.1775968
 6.0    0.1506453
 7.0    0.3000793
 8.0    0.1716508
 9.0   -0.0903336
10.0   -0.2459358
11.0   -0.1711903
12.0    0.0476893
13.0    0.2069261
14.0    0.1710735
15.0   -0.0142245
Bessel Function Y0
15 Values
 0.1   -1.5342387
 1.0    0.0882570
 2.0    0.51037567
 3.0    0.37685001
 4.0   -0.0169407
 5.0   -0.3085176
 6.0   -0.2881947
 7.0   -0.0259497
 8.0    0.2235215
 9.0    0.2499367
10.0    0.0556712
11.0   -0.1688473
12.0   -0.2252373
13.0   -0.0782079
14.0    0.1271926
15.0    0.2054743
Bessel Function J1
20 Values
-5.0    0.3275791
-4.0    0.0660433
-3.0   -0.3390590
-2.0   -0.5767248
-1.0   -0.4400506
 0.0    0.0000000
 1.0    0.4400506
 2.0    0.5767248
 3.0    0.3390590
 4.0   -0.0660433
 5.0   -0.3275791
 6.0   -0.2766839
 7.0   -0.0046828
 8.0    0.2346364
 9.0    0.2453118
10.0    0.0434728
11.0   -0.1767853
12.0   -0.2234471
13.0   -0.0703181
14.0    0.1333752
15.0    0.2051040
Bessel Function Y1
15 Values
 0.1   -6.4589511
```

```
  1.0   -0.7812128
  2.0   -0.1070324
  3.0    0.3246744
  4.0    0.3979257
  5.0    0.1478631
  6.0   -0.1750103
  7.0   -0.3026672
  8.0   -0.1580605
  9.0    0.1043146
 10.0    0.2490154
 11.0    0.1637055
 12.0   -0.0570992
 13.0   -0.2100814
 14.0   -0.1666448
 15.0    0.0210736
Bessel Function Jn, n>=2
20 Values
2       1.0       1.149034849E-01
2       2.0       3.528340286E-01
2       5.0       4.656511628E-02
2       10.0      2.546303137E-01
2       50.0     -5.971280079E-02
5       1.0       2.497577302E-04
5       2.0       7.039629756E-03
5       5.0       2.611405461E-01
5       10.0     -2.340615282E-01
5       50.0     -8.140024770E-02
10      1.0       2.630615124E-10
10      2.0       2.515386283E-07
10      5.0       1.467802647E-03
10      10.0      2.074861066E-01
10      50.0     -1.138478491E-01
20      1.0       3.873503009E-25
20      2.0       3.918972805E-19
20      5.0       2.770330052E-11
20      10.0      1.151336925E-05
20      50.0     -1.167043528E-01
Bessel Function Yn, n>=2
20 Values
2       1.0      -1.650682607
2       2.0      -6.174081042E-01
2       5.0       3.676628826E-01
2       10.0     -5.868082460E-03
2       50.0      9.579316873E-02
5       1.0      -2.604058666E02
5       2.0      -9.935989128
5       5.0      -4.536948225E-01
5       10.0      1.354030477E-01
5       50.0     -7.854841391E-02
10      1.0      -1.216180143E08
10      2.0      -1.291845422E05
10      5.0      -2.512911010E01
10      10.0     -3.598141522E-01
10      50.0      5.723897182E-03
20      1.0      -4.113970315E22
20      2.0      -4.081651389E16
20      5.0      -5.933965297E08
```

```
20      10.0    -1.597483848E03
20      50.0     1.644263395E-02
Modified Bessel Function I0
20 Values
0.0     1.0000000
0.2     1.0100250
0.4     1.0404018
0.6     1.0920453
0.8     1.1665149
1.0     1.2660658
1.2     1.3937256
1.4     1.5533951
1.6     1.7499807
1.8     1.9895593
2.0     2.2795852
2.5     3.2898391
3.0     4.8807925
3.5     7.3782035
4.0     11.301922
4.5     17.481172
5.0     27.239871
6.0     67.234406
8.0     427.56411
10.0    2815.7167
Modified Bessel Function K0
20 Values
0.1     2.4270690
0.2     1.7527038
0.4     1.1145291
0.6     0.77752208
0.8     0.56534710
1.0     0.42102445
1.2     0.31850821
1.4     0.24365506
1.6     0.18795475
1.8     0.14593140
2.0     0.11389387
2.5     6.2347553E-02
3.0     3.4739500E-02
3.5     1.9598897E-02
4.0     1.1159676E-02
4.5     6.3998572E-03
5.0     3.6910983E-03
6.0     1.2439943E-03
8.0     1.4647071E-04
10.0    1.7780062E-05
Modified Bessel Function I1
20 Values
0.0     0.00000000
0.2     0.10050083
0.4     0.20402675
0.6     0.31370403
0.8     0.43286480
1.0     0.56515912
1.2     0.71467794
1.4     0.88609197
1.6     1.0848107
```

```
1.8     1.3171674
2.0     1.5906369
2.5     2.5167163
3.0     3.9533700
3.5     6.2058350
4.0     9.7594652
4.5     15.389221
5.0     24.335643
6.0     61.341937
8.0     399.87313
10.0    2670.9883
Modified Bessel Function K1
20 Values
0.1     9.8538451
0.2     4.7759725
0.4     2.1843544
0.6     1.3028349
0.8     0.86178163
1.0     0.60190724
1.2     0.43459241
1.4     0.32083589
1.6     0.24063392
1.8     0.18262309
2.0     0.13986588
2.5     7.3890816E-02
3.0     4.0156431E-02
3.5     2.2239393E-02
4.0     1.2483499E-02
4.5     7.0780949E-03
5.0     4.0446134E-03
6.0     1.3439197E-03
8.0     1.5536921E-04
10.0    1.8648773E-05
Modified Bessel Function Kn, n>=2
28 Values
2       0.2      49.512430
2       1.0      1.6248389
2       2.0      2.5375975E-01
2       2.5      1.2146021E-01
2       3.0      6.1510459E-02
2       5.0      5.3089437E-03
2       10.0     2.1509817E-05
2       20.0     6.3295437E-10
3       1.0      7.101262825
3       2.0      6.473853909E-01
3       5.0      8.291768415E-03
3       10.0     2.725270026E-05
3       50.0     3.72793677E-23
5       1.0      3.609605896E02
5       2.0      9.431049101
5       5.0      3.270627371E-02
5       10.0     5.754184999E-05
5       50.0     4.36718224E-23
10      1.0      1.807132899E08
10      2.0      1.624824040E05
10      5.0      9.758562829
10      10.0     1.614255300E-03
```

```
10      50.0     9.15098819E-23
20      1.0      6.294369369E22
20      2.0      5.770856853E16
20      5.0      4.827000521E08
20      10.0     1.787442782E02
20      50.0     1.70614838E-21
Modified Bessel Function In, n>=2
28 Values
2       0.2      5.0166876E-03
2       1.0      1.3574767E-01
2       2.0      6.8894844E-01
2       2.5      1.2764661
2       3.0      2.2452125
2       5.0      17.505615
2       10.0     2281.5189
2       20.0     3.9312785E07
3       1.0      2.216842492E-02
3       2.0      2.127399592E-01
3       5.0      1.033115017E01
3       10.0     1.758380717E01
3       50.0     2.67776414E20
5       1.0      2.714631560E-04
5       2.0      9.825679323E-03
5       5.0      2.157974547
5       10.0     7.771882864E02
5       50.0     2.27854831E20
10      1.0      2.752948040E-10
10      2.0      3.016963879E-07
10      5.0      4.580044419E-03
10      10.0     2.189170616E01
10      50.0     1.07159716E20
20      1.0      3.966835986E-25
20      2.0      4.310560576E-19
20      5.0      5.024239358E-11
20      10.0     1.250799736E-04
20      50.0     5.44200840E18
Legendre Polynomials
19 Values
1       0        1.0               1.224745
10      0        1.0               3.240370
20      0        1.0               4.527693
1       0        0.7071067         0.866025
10      0        0.7071067         0.373006
20      0        0.7071067        -0.874140
1       0        0.0               0.000000
10      0        0.0              -0.797435
20      0        0.0               0.797766
2       2        0.7071067         0.484123
10      2        0.7071067        -0.204789
20      2        0.7071067         0.910208
2       2        0.0               0.968246
10      2        0.0               0.804785
20      2        0.0              -0.799672
10      10       0.7071067         0.042505
20      10       0.7071067        -0.707252
10      10       0.0               1.360172
20      10       0.0              -0.853705
```

```
Jacobian Elliptic Function
20 Values
0.0     0.1     0.099833
0.0     0.2     0.19867
0.0     0.5     0.47943
0.0     1.0     0.84147
0.0     2.0     0.90930
0.5     0.1     0.099751
0.5     0.2     0.19802
0.5     0.5     0.47075
0.5     1.0     0.80300
0.5     2.0     0.99466
1.0     0.1     0.099668
1.0     0.2     0.19738
1.0     0.5     0.46212
1.0     1.0     0.76159
1.0     2.0     0.96403
1.0     4.0     0.99933
1.0    -0.2    -0.19738
1.0    -0.5    -0.46212
1.0    -1.0    -0.76159
1.0    -2.0    -0.96403
```

Chapter 7: Random Numbers

Chapter 7 of Numerical Recipes deals with the generation of random numbers drawn from various distributions. The first four subroutines produce uniform deviates with a range of 0.0 to 1.0. `RAN0` *is a subroutine for improving the randomness of a system-supplied random number generator by shuffling the output.* `RAN1` *is a portable random number generator based on three linear congruential generators and a shuffler.* `RAN2` *contains a single congruential generator and a shuffler, and has the advantage of being somewhat faster, if less plentiful in possible output values.* `RAN3` *is another portable generator, based on a subtractive rather than a congruential method.*

The transformation method is used to generate some non-uniform distributions. Resulting from this are routines `EXPDEV`*, which gives exponentially distributed deviates, and* `GASDEV` *for Gaussian deviates. The rejection method of producing non-uniform deviates is also discussed, and is used in* `GAMDEV` *(deviate with a gamma function distribution),* `POIDEV` *(deviate with a Poisson distribution), and* `BNLDEV` *(deviate with a binomial distribution). For generating random sequences of zeros and ones, there are two subroutines,* `IRBIT1` *and* `IRBIT2`*, both based on the 18 lowest significant bits in the seed* `ISEED`*, but each using a different recurrence to proceed from step to step. The national Data Encryption Standard (DES) is discussed as the basis for a random number generator which we call* `RAN4`*. DES itself is carried out by routines* `DES`*,* `KS`*, and* `CYFUN`*.*

* * * *

Sample programs D7R1 to D7R4 are all really the same program, except that each calls a different random number generator (RAN0 to RAN3, respectively). They first draw four consecutive random numbers $X_1, \ldots, X_4$ from the generator in question. Then they treat the numbers as coordinates of a point. For example, they take (X_1, X_2) as a point in two dimensions, (X_1, X_2, X_3) as a point in three dimensions, etc. These points are inside boxes of unit dimension in their respective n-space. They may, however, be either inside or outside of the unit sphere in that space. For $n = 2, 3, 4$ we seek the probability that a point is inside the unit n-sphere. This number is easily calculated theoretically. For $n = 2$ it is $\pi/4$, for $n = 3$ it is $\pi/6$, and for $n = 4$ it is $\pi^2/32$. If the random number generator is not faulty, the points will fall within the unit n-sphere this fraction of the time, and the result should become increasingly accurate as the number of points increases. In programs D7R1 to D7R4 we have taken out a factor of 2^n for convenience, and used the random number generators as a statistical means of determining the value of π, $4\pi/3$, and

$\pi^2/2$.

```
      PROGRAM D7R1
C     Driver for routine RANO
C     Calculates pi statistically using volume of unit n-sphere
      PARAMETER(PI=3.1415926)
      DIMENSION IY(3),YPROB(3)
      FNC(X1,X2,X3,X4)=SQRT(X1**2+X2**2+X3**2+X4**2)
      IDUM=-1
      DO 11 I=1,3
        IY(I)=0
11    CONTINUE
      WRITE(*,'(1X,/,T15,A)') 'Volume of unit n-sphere, n=2,3,4'
      WRITE(*,'(1X,/,T3,A,T17,A,T26,A,T37,A)')
     *        '# points','pi','(4/3)*pi','(1/2)*pi^2'
      DO 14 J=1,15
        DO 12 K=2**(J-1),2**J
          X1=RANO(IDUM)
          X2=RANO(IDUM)
          X3=RANO(IDUM)
          X4=RANO(IDUM)
          IF(FNC(X1,X2,0.0,0.0).LT.1.0) IY(1)=IY(1)+1
          IF(FNC(X1,X2,X3,0.0).LT.1.0) IY(2)=IY(2)+1
          IF(FNC(X1,X2,X3,X4).LT.1.0) IY(3)=IY(3)+1
12      CONTINUE
        DO 13 I=1,3
          YPROB(I)=1.0*(2**(I+1))*IY(I)/(2**J)
13      CONTINUE
        WRITE(*,'(1X,I8,3F12.6)') 2**J,(YPROB(I),I=1,3)
14    CONTINUE
      WRITE(*,'(1X,/,T4,A,3F12.6,/)') 'actual',PI,4.0*PI/3.0,0.5*(PI**2)
      END

      PROGRAM D7R2
C     Driver for routine RAN1
C     Calculates pi statistically using volume of unit n-sphere
      PARAMETER(PI=3.1415926)
      DIMENSION IY(3),YPROB(3)
      FNC(X1,X2,X3,X4)=SQRT(X1**2+X2**2+X3**2+X4**2)
      IDUM=-1
      DO 11 I=1,3
        IY(I)=0
11    CONTINUE
      WRITE(*,'(1X,/,T15,A)') 'Volume of unit n-sphere, n=2,3,4'
      WRITE(*,'(1X,/,T3,A,T17,A,T26,A,T37,A)')
     *        '# points','pi','(4/3)*pi','(1/2)*pi^2'
      DO 14 J=1,15
        DO 12 K=2**(J-1),2**J
          X1=RAN1(IDUM)
          X2=RAN1(IDUM)
          X3=RAN1(IDUM)
          X4=RAN1(IDUM)
          IF(FNC(X1,X2,0.0,0.0).LT.1.0) IY(1)=IY(1)+1
          IF(FNC(X1,X2,X3,0.0).LT.1.0) IY(2)=IY(2)+1
          IF(FNC(X1,X2,X3,X4).LT.1.0) IY(3)=IY(3)+1
12      CONTINUE
        DO 13 I=1,3
```

```
            YPROB(I)=1.0*(2**(I+1))*IY(I)/(2**J)
13        CONTINUE
          WRITE(*,'(1X,I8,3F12.6)') 2**J,(YPROB(I),I=1,3)
14      CONTINUE
        WRITE(*,'(1X,/,T4,A,3F12.6,/)') 'actual',PI,4.0*PI/3.0,0.5*(PI**2)
        END

        PROGRAM D7R3
C       Driver for routine RAN2
C       Calculates pi statistically using volume of unit n-sphere
        PARAMETER(PI=3.1415926)
        DIMENSION IY(3),YPROB(3)
        FNC(X1,X2,X3,X4)=SQRT(X1**2+X2**2+X3**2+X4**2)
        IDUM=-1
        DO 11 I=1,3
          IY(I)=0
11      CONTINUE
        WRITE(*,'(1X,/,T15,A)') 'Volume of unit n-sphere, n=2,3,4'
        WRITE(*,'(1X,/,T3,A,T17,A,T26,A,T37,A)')
     *       '# points','pi','(4/3)*pi','(1/2)*pi^2'
        DO 14 J=1,15
          DO 12 K=2**(J-1),2**J
            X1=RAN2(IDUM)
            X2=RAN2(IDUM)
            X3=RAN2(IDUM)
            X4=RAN2(IDUM)
            IF(FNC(X1,X2,0.0,0.0).LT.1.0) IY(1)=IY(1)+1
            IF(FNC(X1,X2,X3,0.0).LT.1.0) IY(2)=IY(2)+1
            IF(FNC(X1,X2,X3,X4).LT.1.0) IY(3)=IY(3)+1
12        CONTINUE
          DO 13 I=1,3
            YPROB(I)=1.0*(2**(I+1))*IY(I)/(2**J)
13        CONTINUE
          WRITE(*,'(1X,I8,3F12.6)') 2**J,(YPROB(I),I=1,3)
14      CONTINUE
        WRITE(*,'(1X,/,T4,A,3F12.6,/)') 'actual',PI,4.0*PI/3.0,0.5*(PI**2)
        END

        PROGRAM D7R4
C       Driver for routine RAN3
C       Calculates pi statistically using volume of unit n-sphere
        PARAMETER(PI=3.1415926)
        DIMENSION IY(3),YPROB(3)
        FNC(X1,X2,X3,X4)=SQRT(X1**2+X2**2+X3**2+X4**2)
        IDUM=-1
        DO 11 I=1,3
          IY(I)=0
11      CONTINUE
        WRITE(*,'(1X,/,T15,A)') 'Volume of unit n-sphere, n=2,3,4'
        WRITE(*,'(1X,/,T3,A,T17,A,T26,A,T37,A)')
     *       '# points','pi','(4/3)*pi','(1/2)*pi^2'
        DO 14 J=1,15
          DO 12 K=2**(J-1),2**J
            X1=RAN3(IDUM)
            X2=RAN3(IDUM)
            X3=RAN3(IDUM)
            X4=RAN3(IDUM)
```

```
            IF(FNC(X1,X2,0.0,0.0).LT.1.0) IY(1)=IY(1)+1
            IF(FNC(X1,X2,X3,0.0).LT.1.0) IY(2)=IY(2)+1
            IF(FNC(X1,X2,X3,X4).LT.1.0) IY(3)=IY(3)+1
12        CONTINUE
          DO 13 I=1,3
            YPROB(I)=1.0*(2**(I+1))*IY(I)/(2**J)
13        CONTINUE
          WRITE(*,'(1X,I8,3F12.6)') 2**J,(YPROB(I),I=1,3)
14      CONTINUE
        WRITE(*,'(1X,/,T4,A,3F12.6,/)') 'actual',PI,4.0*PI/3.0,0.5*(PI**2)
        END
```

Routine EXPDEV generates random numbers drawn from an exponential deviate. Sample program D7R5 makes ten thousand calls to EXPDEV and bins the results into 21 bins, the contents of which are tallied in array X(I). Then the sum TOTAL of all bins is taken, since some of the numbers will be too large to have fallen in any of the bins. The X(I) are scaled to TOTAL, and then compared to a similarly normalized exponential which is called EXPECT.

```
        PROGRAM D7R5
C       Driver for routine EXPDEV
        PARAMETER(NPTS=10000,EE=2.718281828)
        DIMENSION TRIG(21),X(21)
        DO 11 I=1,21
          TRIG(I)=(I-1)/20.0
          X(I)=0.0
11      CONTINUE
        IDUM=-1
        DO 13 I=1,NPTS
          Y=EXPDEV(IDUM)
          DO 12 J=2,21
            IF((Y.LT.TRIG(J)).AND.(Y.GT.TRIG(J-1))) THEN
              X(J)=X(J)+1.0
            ENDIF
12        CONTINUE
13      CONTINUE
        TOTAL=0.0
        DO 14 I=2,21
          TOTAL=TOTAL+X(I)
14      CONTINUE
        WRITE(*,'(1X,A,I6,A)') 'Exponential distribution with',
     *          NPTS,' points:'
        WRITE(*,'(1X,T5,A,T19,A,T31,A)')
     *          'interval','observed','expected'
        DO 15 I=2,21
          X(I)=X(I)/TOTAL
          EXPECT=EXP(-(TRIG(I-1)+TRIG(I))/2.0)
          EXPECT=EXPECT*0.05*EE/(EE-1)
          WRITE(*,'(1X,2F6.2,2F12.4)')
     *              TRIG(I-1),TRIG(I),X(I),EXPECT
15      CONTINUE
        END
```

GASDEV generates random numbers from a Gaussian deviate. Example D7R6 takes ten thousand of these and puts them into 21 bins. For the purpose of binning, the center of the Gaussian is shifted over by NOVER2=10 bins, to put it in the middle bin. The remainder of the program simply plots the contents of the bins, to illustrate that they have the characteristic Gaussian bell-shape. This allows a quick, though superficial, check of the integrity of the routine.

```
      PROGRAM D7R6
C     Driver for routine GASDEV
      PARAMETER(N=20,NP1=N+1,NOVER2=N/2,NPTS=10000,ISCAL=400,LLEN=50)
      DIMENSION DIST(NP1)
      CHARACTER TEXT(50)*1
      IDUM=-13
      DO 11 J=1,NP1
        DIST(J)=0.0
11    CONTINUE
      DO 12 I=1,NPTS
        J=NINT(0.25*N*GASDEV(IDUM))+NOVER2+1
        IF ((J.GE.1).AND.(J.LE.NP1)) DIST(J)=DIST(J)+1
12    CONTINUE
      WRITE(*,'(1X,A,I6,A)')
     *       'Normally distributed deviate of ',NPTS,' points'
      WRITE(*,'(1X,T6,A,T14,A,T23,A)') 'x','p(x)','graph:'
      DO 15 J=1,NP1
        DIST(J)=DIST(J)/NPTS
        DO 13 K=1,50
          TEXT(K)=' '
13      CONTINUE
        KLIM=INT(ISCAL*DIST(J))
        IF (KLIM.GT.LLEN) KLIM=LLEN
        DO 14 K=1,KLIM
          TEXT(K)='*'
14      CONTINUE
        X=FLOAT(J)/(0.25*N)
        WRITE(*,'(1X,F7.2,F10.4,4X,50A1)')
     *         X,DIST(J),(TEXT(K),K=1,50)
15    CONTINUE
      END
```

The next three sample programs D7R7 to D7R9 are identical to the previous one, but each drives a different random number generator and produces a different graph. D7R7 drives GAMDEV and displays a gamma distribution of order IA specified by the user. D7R8 drives POIDEV and produces a Poisson distribution with mean XM specified by the user. D7R9 drives BNLDEV and produces a binomial distribution, also with specified XM.

```
      PROGRAM D7R7
C     Driver for routine GAMDEV
      PARAMETER(N=20,NPTS=10000,ISCAL=200,LLEN=50)
      DIMENSION DIST(21)
      CHARACTER TEXT(50)*1
      IDUM=-13
10    DO 11 J=1,21
        DIST(J)=0.0
11    CONTINUE
      WRITE(*,*) 'Order of Gamma distribution (n=1..20); -1 to end.'
      READ(*,*) IA
      IF (IA.LE.0) GOTO 99
      IF (IA.GT.20) GOTO 10
      DO 12 I=1,NPTS
```

```
            J=INT(GAMDEV(IA,IDUM))+1
            IF ((J.GE.1).AND.(J.LE.21)) DIST(J)=DIST(J)+1
12        CONTINUE
          WRITE(*,'(1X,A,I2,A,I6,A)') 'Gamma-distribution deviate, order ',
     *          IA,' of ',NPTS,' points'
          WRITE(*,'(1X,T6,A,T14,A,T23,A)') 'x','p(x)','graph:'
          DO 15 J=1,20
            DIST(J)=DIST(J)/NPTS
            DO 13 K=1,50
              TEXT(K)=' '
13          CONTINUE
            KLIM=INT(ISCAL*DIST(J))
            IF (KLIM.GT.LLEN) KLIM=LLEN
            DO 14 K=1,KLIM
              TEXT(K)='*'
14          CONTINUE
            WRITE(*,'(1X,F7.2,F10.4,4X,50A1)')
     *              FLOAT(J),DIST(J),(TEXT(K),K=1,50)
15        CONTINUE
          GOTO 10
99        END

          PROGRAM D7R8
C         Driver for routine POIDEV
          PARAMETER(N=20,NPTS=10000,ISCAL=200,LLEN=50)
          DIMENSION DIST(21)
          CHARACTER TEXT(50)*1
          IDUM=-13
10        DO 11 J=1,21
            DIST(J)=0.0
11        CONTINUE
          WRITE(*,*) 'Mean of Poisson distrib. (x=0 to 20); neg. to end'
          READ(*,*) XM
          IF (XM.LT.0.0) GOTO 99
          IF (IA.GT.20.0) GOTO 10
          DO 12 I=1,NPTS
            J=INT(POIDEV(XM,IDUM))+1
            IF ((J.GE.1).AND.(J.LE.21)) DIST(J)=DIST(J)+1
12        CONTINUE
          WRITE(*,'(1X,A,F5.2,A,I6,A)')
     *          'Poisson-distributed deviate, mean ',
     *          XM,' of ',NPTS,' points'
          WRITE(*,'(1X,T6,A,T14,A,T23,A)') 'x','p(x)','graph:'
          DO 15 J=1,20
            DIST(J)=DIST(J)/NPTS
            DO 13 K=1,50
              TEXT(K)=' '
13          CONTINUE
            KLIM=INT(ISCAL*DIST(J))
            IF (KLIM.GT.LLEN) KLIM=LLEN
            DO 14 K=1,KLIM
              TEXT(K)='*'
14          CONTINUE
            WRITE(*,'(1X,F7.2,F10.4,4X,50A1)')
     *              FLOAT(J),DIST(J),(TEXT(K),K=1,50)
15        CONTINUE
          GOTO 10
```

```
99      END

        PROGRAM D7R9
C       Driver for routine BNLDEV
        PARAMETER(N=20,NPTS=1000,ISCAL=200,NN=100)
        DIMENSION DIST(21)
        CHARACTER TEXT(50)*1
        IDUM=-133
        LLEN=50
10      DO 11 J=1,21
            DIST(J)=0.0
11      CONTINUE
        WRITE(*,*) 'Mean of binomial distribution (0 to 20) (neg to end)'
        READ(*,*) XM
        IF (XM.LT.0) GOTO 99
        PP=XM/NN
        DO 12 I=1,NPTS
            J=INT(BNLDEV(PP,NN,IDUM))
            IF ((J.GE.0).AND.(J.LE.20)) DIST(J+1)=DIST(J+1)+1
12      CONTINUE
        WRITE(*,'(1X,T5,A,T10,A,T18,A)') 'x','p(x)','graph:'
        DO 15 J=1,20
            DIST(J)=DIST(J)/NPTS
            DO 13 K=1,50
                TEXT(K)=' '
13          CONTINUE
            TEXT(1)='*'
            KLIM=INT(ISCAL*DIST(J))
            IF (KLIM.GT.LLEN) KLIM=LLEN
            DO 14 K=1,KLIM
                TEXT(K)='*'
14          CONTINUE
            WRITE(*,'(1X,F5.1,F8.4,3X,50A1)') FLOAT(J-1),DIST(J),
     *              (TEXT(K),K=1,50)
15      CONTINUE
        GOTO 10
99      END
```

Subroutines IRBIT1 and IRBIT2 both generate random series of ones and zeros. The sample programs D7R10 and D7R11 for the two are the same, and they check that the series have correct statistical properties (or more exactly, that they have at least one correct property). They look for a 1 in the series and count how many zeros follow it before the next 1 appears. The result is stored as a distribution. There should be, for example, a 50% chance of no zeros, a 25% chance of exactly one zero, and so on.

```
        PROGRAM D7R10
C       Driver for routine IRBIT1
C       Calculate distribution of runs of zeros
        PARAMETER(NBIN=15,NTRIES=10000)
        DIMENSION DELAY(NBIN)
        ISEED=12345
        DO 11 I=1,NBIN
            DELAY(I)=0.0
11      CONTINUE
        IPTS=0
```

```
      DO 13 I=1,NTRIES
        IF (IRBIT1(ISEED).EQ.1) THEN
          IPTS=IPTS+1
          IFLG=0
          DO 12 J=1,NBIN
            IF ((IRBIT1(ISEED).EQ.1)
     *          .AND.(IFLG.EQ.0)) THEN
              IFLG=1
              DELAY(J)=DELAY(J)+1.0
            ENDIF
12        CONTINUE
        ENDIF
13    CONTINUE
      WRITE(*,*) 'Distribution of runs of N zeros'
      WRITE(*,'(1X,T7,A,T16,A,T38,A)') 'N','Probability','Expected'
      DO 14 N=1,NBIN
        WRITE(*,'(1X,I6,F18.6,F20.6)')
     *          N-1,DELAY(N)/IPTS,1/(2.0**N)
14    CONTINUE
      END

      PROGRAM D7R11
C     Driver for routine IRBIT2
C     Calculate distribution of runs of zeros
      PARAMETER(NBIN=15,NTRIES=10000)
      DIMENSION DELAY(NBIN)
      ISEED=111
      DO 11 I=1,NBIN
        DELAY(I)=0.0
11    CONTINUE
      IPTS=0
      DO 13 I=1,NTRIES
        IF (IRBIT2(ISEED).EQ.1) THEN
          IPTS=IPTS+1
          IFLG=0
          DO 12 J=1,NBIN
            IF ((IRBIT2(ISEED).EQ.1)
     *          .AND.(IFLG.EQ.0)) THEN
              IFLG=1
              DELAY(J)=DELAY(J)+1.0
            ENDIF
12        CONTINUE
        ENDIF
13    CONTINUE
      WRITE(*,*) 'Distribution of runs of N zeros'
      WRITE(*,'(1X,T7,A,T16,A,T38,A)') 'N','Probability','Expected'
      DO 14 N=1,NBIN
        WRITE(*,'(1X,I6,F18.6,F20.6)')
     *          N-1,DELAY(N)/IPTS,1/(2.0**N)
14    CONTINUE
      END
```

The next routine, RAN4, is a random number generator with a uniform deviate, based on the data encryption standard DES. When applied to RAN4, the routine we used to demonstrate RAN0 to RAN3 is outrageously time consuming. RAN4 is random but very slow. To try out RAN4 we simply list the first ten random numbers for a given seed IDUM=-123. Compare your results to these; they should be exactly the same. We also generate 50 more random numbers and find their average and variance with AVEVAR.

```
      PROGRAM D7R12
C     Driver for routine RAN4
      PARAMETER(NPT=50)
      DIMENSION Y(NPT)
      IDUM=-123
      AVE=0.0
      WRITE(*,'(1X,A,I5/)') 'First 10 random numbers with IDUM =',IDUM
      WRITE(*,'(1X,A4,A11)') '#','RAN4'
      DO 11 J=1,10
        WRITE(*,'(1X,I4,F12.6)') J,RAN4(IDUM)
11    CONTINUE
      WRITE(*,'(/1X,A,I3)') 'Average and Variance of next ',NPT
      DO 12 J=1,NPT
        Y(J)=RAN4(IDUM)
12    CONTINUE
      CALL AVEVAR(Y,NPT,AVE,VAR)
      WRITE(*,'(/1X,A,F10.4)') 'Average: ',AVE
      WRITE(*,'(1X,A,F10.4/)') 'Variance:',VAR
      WRITE(*,'(/1X,A)') 'Expected Result for an Infinite Sample:'
      WRITE(*,'(/1X,A,F10.4)') 'Average: ',0.5
      WRITE(*,'(1X,A,F10.4/)') 'Variance:',1./12.
      END
```

```
 First 10 random numbers with IDUM = -123
         1      .076597
         2      .533635
         3      .919756
         4      .317618
         5      .187471
         6      .629516
         7      .588766
         8      .953446
         9      .366207
         0      .915449
 Average and Variance of next  50
         Average:       .4542
         Variance:      .0786
```

DES is a software implementation of the national data encryption standard. The complete formal test for this standard, though long and detailed, is included in program D7R13. This test consists of feeding in a long series of input codes and checking the output for agreement with expected output codes. The input-output pairs comprising the test are contained in file DESTST.DAT, listed in the Appendix to this chapter. To save you the job of comparing the many 16-character strings for accuracy, we have ended each line with the phrase "o.k." or "wrong", depending on the outcome.

```
      PROGRAM D7R13
C     Driver for routine DES
      DIMENSION IN(64),KEY(64),IOUT(64),ICMP(64)
      INTEGER HEX2IN
      CHARACTER HIN(17)*1,HKEY(17)*1,HOUT(17)*1,HCMP(17)*1,
```

```
     *             VERDCT*8,TEXT*60,TEXT2*6,IN2HEX*1
      OPEN(5,FILE='DESTST.DAT',STATUS='OLD')
      READ (5,'(A)') TEXT
      WRITE(*,'(/1X,A)') TEXT
5     READ (5,'(A)') TEXT
      WRITE(*,'(/1X,A,/)') TEXT
      READ (5,*,ERR=99) NCIPHR
      READ (5,'(A)') TEXT2
      IF (TEXT2.EQ.'encode') IDIREC=0
      IF (TEXT2.EQ.'decode') IDIREC=1
10    WRITE(*,'(1X,T9,A,T23,A,T38,A,T56,A)')
     *        'Key','Plaintext','Expected Cipher','Actual Cipher'
      MM=MIN(NCIPHR,16)
      NCIPHR=NCIPHR-16
      DO 16 M=1,MM
        READ(5,'(51A1)') (HKEY(K),K=1,17),(HIN(K),K=1,17),
     *          (HCMP(K),K=1,17)
        DO 12 I=1,16
          J=I+1
          IDUM=HEX2IN(HIN(J))
          JDUM=HEX2IN(HKEY(J))
          DO 11 K=1,4
            L=4*I+1-K
            IN(L)=MOD(IDUM,2)
            IDUM=IDUM/2
            KEY(L)=MOD(JDUM,2)
            JDUM=JDUM/2
11        CONTINUE
12      CONTINUE
        NEWKEY=1
        CALL DES(IN,KEY,NEWKEY,IDIREC,IOUT)
        HOUT(1)=' '
        DO 14 I=1,16
          JDUM=0
          DO 13 J=1,4
            JDUM=JDUM+(2**(4-J))*IOUT(4*(I-1)+J)
13        CONTINUE
          HOUT(I+1)=IN2HEX(JDUM)
14      CONTINUE
        VERDCT='    o.k.'
        DO 15 I=1,17
          IF (HCMP(I).NE.HOUT(I)) VERDCT='   wrong'
15      CONTINUE
        WRITE(*,'(1X,68A1,A)') (HKEY(K),K=1,17),(HIN(K),K=1,17),
     *          (HCMP(K),K=1,17),(HOUT(K),K=1,17),VERDCT
16    CONTINUE
      WRITE(*,'(/1X,A)') 'press RETURN to continue...'
      READ(*,*)
      IF (NCIPHR.LE.0) GOTO 5
      GOTO 10
99    CLOSE(5)
      END
      INTEGER FUNCTION HEX2IN(CH)
C     Converts character representing hexadecimal data to its integer
C     value in a machine-independent way
      CHARACTER CH*1
      IF(CH.GE.'0'.AND.CH.LE.'9')THEN
```

```
            HEX2IN=ICHAR(CH)-ICHAR('0')
         ELSE
            HEX2IN=ICHAR(CH)-ICHAR('A')+10
         ENDIF
         RETURN
         END
         CHARACTER*1 FUNCTION IN2HEX(I)
C        Inverse of HEX2IN
         IF(I.LE.9)THEN
            IN2HEX=CHAR(I+ICHAR('0'))
         ELSE
            IN2HEX=CHAR(I-10+ICHAR('A'))
         ENDIF
         RETURN
         END
```

DESKS contains two auxiliary routines for DES, namely CYFUN and KS. Technically, these were fully tested by the previous formal procedure. Sample program D7R14 is simply an additional routine to help you track down problems. It first feeds a single KEY to KS and generates 16 subkeys. These subkeys are strings of zeros and ones which we print out as strings of "-" and "*". In this form the results are easier to compare with printed text. Next an input vector is fed to the cipher function CYFUN along with each of these subkeys. The results are again printed in "-" and "*" characters. If all is well, you will observe these patterns:

```
Legend:
                       -=0    *=1
Master key:
      *-*-*-*-*-*-*-*-*-*-*-*-*-*-*-*-*-*-*-*-*-*-*-*-*-*-*-*-
Sub-master keys:
    1     -*--****-**-**-*--**-*-*--*-*-*-*****-**-*-*-**
    2     -*--*******--*-*--*-**-*--*-***--*****-**-*-*-**
    3     **--*-***----*-**-*-*****-*-***--*-**--*-***--**
    4     *****--**---*-*-*-*-****---*****--*-**-***--*-
    5     *-**--**-***-*-*-*-*-**-*-*-**---****-*-**---
    6     *-**-----****-**-*-**-**-**--**--*-**--*-***--
    7     -***-*---******--*-*-*---*-**--*****-**-*-*-**--
    8     -*---**-****-*-*-***-*----***----*****--*-*-**-*
    9     **---**-***--*-*-***-*-*--***-*--*****--*-**-*-*
   10     **--******---***--**--***-*-*-**-**-*--**-**--**
   11     ***-*****--*--***-*-*-***-*--***-**-*-**---*--**
   12     *-***-***--*--*-**--*-******-***------**-*-*-**-
   13     --***--*-*-**-*-**-**-*-**-*-*-**-----****--***-
   14     --**-*---****-***-***---*-*-*--*-**-**-**--**-*
   15     ---*-**--**-**-*-*-*-*-*-****-*-*-**-*--***-**-*
   16     -*-****--**-**-*-*-*-*-*-**-*-*-****-*--*-*-****
Legend:
                       -=0    *=1
Input to cipher function:
      *--*--*--*--*--*--*--*--*--*--*-
Ciphered output:
    1     -*----**-*-------****-----***-*
    2     --*--***-**--*-*--****-*-*---*-*
    3     --*****--*-*--*-**---*----***--*
    4     -**-*--*---*-**---***-****---*-*
    5     *-*---*---*****--*-*-*-*--*-**--
```

```
      6    --*--**--*---**--*-**----*-----*
      7    *-*-****-*-**-*------*-*----****
      8    -***--***----**-*-----*---*--**-
      9    -**-*-*****-***-****---------*--
     10    *--***--**-******-*--**-*-***-*-
     11    -*****-*----**--******-*-**-*---
     12    ***-*-----***---**---*--*-*-*-*-
     13    -**-*-----*-****-*----------**-*
     14    ----*---**-***----*-*---**------
     15    *-**--*-*-**--**-*-*--***-*-****
     16    *---*--*---*-********-***-*-***-
```

```
      PROGRAM D7R14
C     Driver for routines KS and CYFUN in file DESKS.FOR
      DIMENSION KEY(64),KN(48),IR(32),IOUT(32)
      CHARACTER TEXT(64)*1
C     First test routine KS
      DO 11 I=1,64
        KEY(I)=MOD(I,2)
        IF (KEY(I).EQ.0) TEXT(I)='-'
        IF (KEY(I).EQ.1) TEXT(I)='*'
11    CONTINUE
      WRITE(*,*) 'Legend:'
      WRITE(*,'(1X,T12,A)') '-=0   *=1 '
      WRITE(*,*) 'Master key:'
      WRITE(*,'(1X,T12,56A1)') (TEXT(I),I=1,56)
      WRITE(*,*) 'Sub-master keys:'
      DO 13 I=1,16
        CALL KS(KEY,I,KN)
        DO 12 K=1,48
          IF (KN(K).EQ.0) TEXT(K)='-'
          IF (KN(K).EQ.1) TEXT(K)='*'
12      CONTINUE
        WRITE(*,'(1X,I6,T12,48A1)') I,(TEXT(J),J=1,48)
13    CONTINUE
      WRITE(*,*) 'press RETURN to continue...'
      READ(*,*)
C     Now test routine CYFUN
      DO 14 I=1,32
        IR(I)=MOD(I,3)
        IR(I)=MOD(IR(I),2)
        IF (IR(I).EQ.0) TEXT(I)='-'
        IF (IR(I).EQ.1) TEXT(I)='*'
14    CONTINUE
      WRITE(*,*) 'Legend:'
      WRITE(*,'(1X,T12,A)') '-=0   *=1 '
      WRITE(*,*) 'Input to cipher function:'
      WRITE(*,'(1X,T12,32A1)') (TEXT(I),I=1,32)
      WRITE(*,*) 'Ciphered output:'
      DO 16 I=1,16
        CALL KS(KEY,I,KN)
        CALL CYFUN(IR,KN,IOUT)
        DO 15 K=1,32
          IF (IOUT(K).EQ.0) TEXT(K)='-'
          IF (IOUT(K).EQ.1) TEXT(K)='*'
15      CONTINUE
        WRITE(*,'(1X,I6,T12,32A1)') I,(TEXT(J),J=1,32)
```

```
16      CONTINUE
        END
```

Appendix

File DESTST.DAT:

```
DES Validation, as per NBS publication 500-20
*** Initial Permutation and Expansion test: ***
       Key            Plaintext     Expected Cipher
 0101010101010101 95F8A5E5DD31D900 8000000000000000
 0101010101010101 DD7F121CA5015619 4000000000000000
 0101010101010101 2E8653104F3834EA 2000000000000000
 0101010101010101 4BD388FF6CD81D4F 1000000000000000
 0101010101010101 20B9E767B2FB1456 0800000000000000
 0101010101010101 55579380D77138EF 0400000000000000
 0101010101010101 6CC5DEFAAF04512F 0200000000000000
 0101010101010101 0D9F279BA5D87260 0100000000000000
 0101010101010101 D9031B0271BD5A0A 0080000000000000
 0101010101010101 424250B37C3DD951 0040000000000000
 0101010101010101 B8061B7ECD9A21E5 0020000000000000
 0101010101010101 F15D0F286B65BD28 0010000000000000
 0101010101010101 ADD0CC8D6E5DEBA1 0008000000000000
 0101010101010101 E6D5F82752AD63D1 0004000000000000
 0101010101010101 ECBFE3BD3F591A5E 0002000000000000
 0101010101010101 F356834379D165CD 0001000000000000
 0101010101010101 2B9F982F20037FA9 0000800000000000
 0101010101010101 889DE068A16F0BE6 0000400000000000
 0101010101010101 E19E275D846A1298 0000200000000000
 0101010101010101 329A8ED523D71AEC 0000100000000000
 0101010101010101 E7FCE22557D23C97 0000080000000000
 0101010101010101 12A9F5817FF2D65D 0000040000000000
 0101010101010101 A484C3AD38DC9C19 0000020000000000
 0101010101010101 FBE00A8A1EF8AD72 0000010000000000
 0101010101010101 750D079407521363 0000008000000000
 0101010101010101 64FEED9C724C2FAF 0000004000000000
 0101010101010101 F02B263B328E2B60 0000002000000000
 0101010101010101 9D64555A9A10B852 0000001000000000
 0101010101010101 D106FF0BED5255D7 0000000800000000
 0101010101010101 E1652C6B138C64A5 0000000400000000
 0101010101010101 E428581186EC8F46 0000000200000000
 0101010101010101 AEB5F5EDE22D1A36 0000000100000000
 0101010101010101 E943D7568AEC0C5C 0000000080000000
 0101010101010101 DF98C8276F54B04B 0000000040000000
 0101010101010101 B160E4680F6C696F 0000000020000000
 0101010101010101 FA0752B07D9C4AB8 0000000010000000
 0101010101010101 CA3A2B036DBC8502 0000000008000000
 0101010101010101 5E0905517BB59BCF 0000000004000000
 0101010101010101 814EEB3B91D90726 0000000002000000
 0101010101010101 4D49DB1532919C9F 0000000001000000
 0101010101010101 25EB5FC3F8CF0621 0000000000800000
 0101010101010101 AB6A20C0620D1C6F 0000000000400000
 0101010101010101 79E90DBC98F92CCA 0000000000200000
 0101010101010101 866ECEDD8072BB0E 0000000000100000
 0101010101010101 8B54536F2F3E64A8 0000000000080000
 0101010101010101 EA51D3975595B86B 0000000000040000
 0101010101010101 CAFFC6AC4542DE31 0000000000020000
```

```
0101010101010101 8DD45A2DDF90796C 0000000000010000
0101010101010101 1029D55E880EC2D0 0000000000008000
0101010101010101 5D86CB23639DBEA9 0000000000004000
0101010101010101 1D1CA853AE7C0C5F 0000000000002000
0101010101010101 CE332329248F3228 0000000000001000
0101010101010101 8405D1ABE24FB942 0000000000000800
0101010101010101 E643D78090CA4207 0000000000000400
0101010101010101 48221B9937748A23 0000000000000200
0101010101010101 DD7C0BBD61FAFD54 0000000000000100
0101010101010101 2FBC291A570DB5C4 0000000000000080
0101010101010101 E07C30D7E4E26E12 0000000000000040
0101010101010101 0953E2258E8E90A1 0000000000000020
0101010101010101 5B711BC4CEEBF2EE 0000000000000010
0101010101010101 CC083F1E6D9E85F6 0000000000000008
0101010101010101 D2FD8867D50D2DFE 0000000000000004
0101010101010101 06E7EA22CE92708F 0000000000000002
0101010101010101 166B40B44ABA4BD6 0000000000000001
*** Inverse Permutation and Expansion test ***
      Key             Plaintext       Expected Cipher
0101010101010101 8000000000000000 95F8A5E5DD31D900
0101010101010101 4000000000000000 DD7F121CA5015619
0101010101010101 2000000000000000 2E8653104F3834EA
0101010101010101 1000000000000000 4BD388FF6CD81D4F
0101010101010101 0800000000000000 20B9E767B2FB1456
0101010101010101 0400000000000000 55579380D77138EF
0101010101010101 0200000000000000 6CC5DEFAAF04512F
0101010101010101 0100000000000000 0D9F279BA5D87260
0101010101010101 0080000000000000 D9031B0271BD5A0A
0101010101010101 0040000000000000 424250B37C3DD951
0101010101010101 0020000000000000 B8061B7ECD9A21E5
0101010101010101 0010000000000000 F15D0F286B65BD28
0101010101010101 0008000000000000 ADD0CC8D6E5DEBA1
0101010101010101 0004000000000000 E6D5F82752AD63D1
0101010101010101 0002000000000000 ECBFE3BD3F591A5E
0101010101010101 0001000000000000 F356834379D165CD
0101010101010101 0000800000000000 2B9F982F20037FA9
0101010101010101 0000400000000000 889DE068A16F0BE6
0101010101010101 0000200000000000 E19E275D846A1298
0101010101010101 0000100000000000 329A8ED523D71AEC
0101010101010101 0000080000000000 E7FCE22557D23C97
0101010101010101 0000040000000000 12A9F5817FF2D65D
0101010101010101 0000020000000000 A484C3AD38DC9C19
0101010101010101 0000010000000000 FBE00A8A1EF8AD72
0101010101010101 0000008000000000 750D079407521363
0101010101010101 0000004000000000 64FEED9C724C2FAF
0101010101010101 0000002000000000 F02B263B328E2B60
0101010101010101 0000001000000000 9D64555A9A10B852
0101010101010101 0000000800000000 D106FF0BED5255D7
0101010101010101 0000000400000000 E1652C6B138C64A5
0101010101010101 0000000200000000 E428581186EC8F46
0101010101010101 0000000100000000 AEB5F5EDE22D1A36
0101010101010101 0000000080000000 E943D7568AEC0C5C
0101010101010101 0000000040000000 DF98C8276F54B04B
0101010101010101 0000000020000000 B160E4680F6C696F
0101010101010101 0000000010000000 FA0752B07D9C4AB8
0101010101010101 0000000008000000 CA3A2B036DBC8502
0101010101010101 0000000004000000 5E0905517BB59BCF
```

```
0101010101010101 0000000002000000 814EEB3B91D90726
0101010101010101 0000000001000000 4D49DB1532919C9F
0101010101010101 0000000000800000 25EB5FC3F8CF0621
0101010101010101 0000000000400000 AB6A20C0620D1C6F
0101010101010101 0000000000200000 79E90DBC98F92CCA
0101010101010101 0000000000100000 866ECEDD8072BB0E
0101010101010101 0000000000080000 8B54536F2F3E64A8
0101010101010101 0000000000040000 EA51D3975595B86B
0101010101010101 0000000000020000 CAFFC6AC4542DE31
0101010101010101 0000000000010000 8DD45A2DDF90796C
0101010101010101 0000000000008000 1029D55E880EC2D0
0101010101010101 0000000000004000 5D86CB23639DBEA9
0101010101010101 0000000000002000 1D1CA853AE7C0C5F
0101010101010101 0000000000001000 CE332329248F3228
0101010101010101 0000000000000800 8405D1ABE24FB942
0101010101010101 0000000000000400 E643D78090CA4207
0101010101010101 0000000000000200 48221B9937748A23
0101010101010101 0000000000000100 DD7C0BBD61FAFD54
0101010101010101 0000000000000080 2FBC291A570DB5C4
0101010101010101 0000000000000040 E07C30D7E4E26E12
0101010101010101 0000000000000020 0953E2258E8E90A1
0101010101010101 0000000000000010 5B711BC4CEEBF2EE
0101010101010101 0000000000000008 CC083F1E6D9E85F6
0101010101010101 0000000000000004 D2FD8867D50D2DFE
0101010101010101 0000000000000002 06E7EA22CE92708F
0101010101010101 0000000000000001 166B40B44ABA4BD6
*** Key Permutation tests: ***
       Key           Plaintext        Expected Cipher
8001010101010101 0000000000000000 95A8D72813DAA94D
4001010101010101 0000000000000000 0EEC1487DD8C26D5
2001010101010101 0000000000000000 7AD16FFB79C45926
1001010101010101 0000000000000000 D3746294CA6A6CF3
0801010101010101 0000000000000000 809F5F873C1FD761
0401010101010101 0000000000000000 C02FAFFEC989D1FC
0201010101010101 0000000000000000 4615AA1D33E72F10
0180010101010101 0000000000000000 2055123350C00858
0140010101010101 0000000000000000 DF3B99D6577397C8
0120010101010101 0000000000000000 31FE17369B5288C9
0110010101010101 0000000000000000 DFDD3CC64DAE1642
0108010101010101 0000000000000000 178C83CE2B399D94
0104010101010101 0000000000000000 50F636324A9B7F80
0102010101010101 0000000000000000 A8468EE3BC18F06D
0101800101010101 0000000000000000 A2DC9E92FD3CDE92
0101400101010101 0000000000000000 CAC09F797D031287
0101200101010101 0000000000000000 90BA680B22AEB525
0101100101010101 0000000000000000 CE7A24F350E280B6
0101080101010101 0000000000000000 882BFF0AA01A0B87
0101040101010101 0000000000000000 25610288924511C2
0101020101010101 0000000000000000 C71516C29C75D170
0101018001010101 0000000000000000 5199C29A52C9F059
0101014001010101 0000000000000000 C22F0A294A71F29F
0101012001010101 0000000000000000 EE371483714C02EA
0101011001010101 0000000000000000 A81FBD448F9E522F
0101010801010101 0000000000000000 4F644C92E192DFED
0101010401010101 0000000000000000 1AFA9A66A6DF92AE
0101010201010101 0000000000000000 B3C1CC715CB879D8
0101010180010101 0000000000000000 19D032E64AB0BD8B
```

```
0101010140010101 0000000000000000 3CFAA7A7DC8720DC
0101010120010101 0000000000000000 B7265F7F447AC6F3
0101010110010101 0000000000000000 9DB73B3C0D163F54
0101010108010101 0000000000000000 8181B65BABF4A975
0101010104010101 0000000000000000 93C9B64042EAA240
0101010102010101 0000000000000000 5570530829705592
0101010101800101 0000000000000000 8638809E878787A0
0101010101400101 0000000000000000 41B9A79AF79AC208
0101010101200101 0000000000000000 7A9BE42F2009A892
0101010101100101 0000000000000000 29038D56BA6D2745
0101010101080101 0000000000000000 5495C6ABF1E5DF51
0101010101040101 0000000000000000 AE13DBD561488933
0101010101020101 0000000000000000 024D1FFA8904E389
0101010101018001 0000000000000000 D1399712F99BF02E
0101010101014001 0000000000000000 14C1D7C1CFFEC79E
0101010101012001 0000000000000000 1DE5279DAE3BED6F
0101010101011001 0000000000000000 E941A33F85501303
0101010101010801 0000000000000000 DA99DBBC9A03F379
0101010101010401 0000000000000000 B7FC92F91D8E92E9
0101010101010201 0000000000000000 AE8E5CAA3CA04E85
0101010101010180 0000000000000000 9CC62DF43B6EED74
0101010101010140 0000000000000000 D863DBB5C59A91A0
0101010101010120 0000000000000000 A1AB2190545B91D7
0101010101010110 0000000000000000 0875041E64C570F7
0101010101010108 0000000000000000 5A594528BEBEF1CC
0101010101010104 0000000000000000 FCDB3291DE21F0C0
0101010101010102 0000000000000000 869EFD7F9F265A09
*** Test of right-shifts in Decryption ***
      Key             Plaintext       Expected Cipher
8001010101010101 95A8D72813DAA94D 0000000000000000
4001010101010101 0EEC1487DD8C26D5 0000000000000000
2001010101010101 7AD16FFB79C45926 0000000000000000
1001010101010101 D3746294CA6A6CF3 0000000000000000
0801010101010101 809F5F873C1FD761 0000000000000000
0401010101010101 C02FAFFEC989D1FC 0000000000000000
0201010101010101 4615AA1D33E72F10 0000000000000000
0180010101010101 2055123350C00858 0000000000000000
0140010101010101 DF3B99D6577397C8 0000000000000000
0120010101010101 31FE17369B5288C9 0000000000000000
0110010101010101 DFDD3CC64DAE1642 0000000000000000
0108010101010101 178C83CE2B399D94 0000000000000000
0104010101010101 50F636324A9B7F80 0000000000000000
0102010101010101 A8468EE3BC18F06D 0000000000000000
0101800101010101 A2DC9E92FD3CDE92 0000000000000000
0101400101010101 CAC09F797D031287 0000000000000000
0101200101010101 90BA680B22AEB525 0000000000000000
0101100101010101 CE7A24F350E280B6 0000000000000000
0101080101010101 882BFF0AA01A0B87 0000000000000000
0101040101010101 25610288924511C2 0000000000000000
0101020101010101 C71516C29C75D170 0000000000000000
0101018001010101 5199C29A52C9F059 0000000000000000
0101014001010101 C22F0A294A71F29F 0000000000000000
0101012001010101 EE371483714C02EA 0000000000000000
0101011001010101 A81FBD448F9E522F 0000000000000000
0101010801010101 4F644C92E192DFED 0000000000000000
0101010401010101 1AFA9A66A6DF92AE 0000000000000000
0101010201010101 B3C1CC715CB879D8 0000000000000000
```

```
0101010180010101 19D032E64AB0BD8B 0000000000000000
0101010140010101 3CFAA7A7DC8720DC 0000000000000000
0101010120010101 B7265F7F447AC6F3 0000000000000000
0101010110010101 9DB73B3C0D163F54 0000000000000000
0101010108010101 8181B65BABF4A975 0000000000000000
0101010104010101 93C9B64042EAA240 0000000000000000
0101010102010101 5570530829705592 0000000000000000
0101010101800101 8638809E878787A0 0000000000000000
0101010101400101 41B9A79AF79AC208 0000000000000000
0101010101200101 7A9BE42F2009A892 0000000000000000
0101010101100101 29038D56BA6D2745 0000000000000000
0101010101080101 5495C6ABF1E5DF51 0000000000000000
0101010101040101 AE13DBD561488933 0000000000000000
0101010101020101 024D1FFA8904E389 0000000000000000
0101010101018001 D1399712F99BF02E 0000000000000000
0101010101014001 14C1D7C1CFFEC79E 0000000000000000
0101010101012001 1DE5279DAE3BED6F 0000000000000000
0101010101011001 E941A33F85501303 0000000000000000
0101010101010801 DA99DBBC9A03F379 0000000000000000
0101010101010401 B7FC92F91D8E92E9 0000000000000000
0101010101010201 AE8E5CAA3CA04E85 0000000000000000
0101010101010180 9CC62DF43B6EED74 0000000000000000
0101010101010140 D863DBB5C59A91A0 0000000000000000
0101010101010120 A1AB2190545B91D7 0000000000000000
0101010101010110 0875041E64C570F7 0000000000000000
0101010101010108 5A594528BEBEF1CC 0000000000000000
0101010101010104 FCDB3291DE21F0C0 0000000000000000
0101010101010102 869EFD7F9F265A09 0000000000000000
*** Data permutation test: ***
      Key              Plaintext        Expected Cipher
1046913489980131 0000000000000000 88D55E54F54C97B4
1007103489988020 0000000000000000 0C0CC00C83EA48FD
10071034C8980120 0000000000000000 83BC8EF3A6570183
1046103489988020 0000000000000000 DF725DCAD94EA2E9
1086911519190101 0000000000000000 E652B53B550BE8B0
1086911519580101 0000000000000000 AF527120C485CBB0
5107B01519580101 0000000000000000 0F04CE393DB926D5
1007B01519190101 0000000000000000 C9F00FFC74079067
3107915498080101 0000000000000000 7CFD82A593252B4E
3107919498080101 0000000000000000 CB49A2F9E91363E3
10079115B9080140 0000000000000000 00B588BE70D23F56
3107911598080140 0000000000000000 406A9A6AB43399AE
1007D01589980101 0000000000000000 6CB773611DCA9ADA
9107911589980101 0000000000000000 67FD21C17DBB5D70
9107D01589190101 0000000000000000 9592CB4110430787
1007D01598980120 0000000000000000 A6B7FF68A318DDD3
1007940498190101 0000000000000000 4D102196C914CA16
0107910491190401 0000000000000000 2DFA9F4573594965
0107910491190101 0000000000000000 B46604816C0E0774
0107940491190401 0000000000000000 6E7E6221A4F34E87
19079210981A0101 0000000000000000 AA85E74643233199
1007911998190801 0000000000000000 2E5A19DB4D1962D6
10079119981A0801 0000000000000000 23A866A809D30894
1007921098190101 0000000000000000 D812D961F017D320
100791159819010B 0000000000000000 055605816E58608F
1004801598190101 0000000000000000 ABD88E8B1B7716F1
1004801598190102 0000000000000000 537AC95BE69DA1E1
```

```
 1004801598190108 0000000000000000 AED0F6AE3C25CDD8
 1002911498100104 0000000000000000 B3E35A5EE53E7B8D
 1002911598190104 0000000000000000 61C79C71921A2EF8
 1002911598100201 0000000000000000 E2F5728F0995013C
 1002911698100101 0000000000000000 1AEAC39A61F0A464
*** S-Box test: ***
       Key             Plaintext       Expected Cipher
 7CA110454A1A6E57 01A1D6D039776742 690F5B0D9A26939B
 0131D9619DC1376E 5CD54CA83DEF57DA 7A389D10354BD271
 07A1133E4A0B2686 0248D43806F67172 868EBB51CAB4599A
 3849674C2602319E 51454B582DDF440A 7178876E01F19B2A
 04B915BA43FEB5B6 42FD443059577FA2 AF37FB421F8C4095
 0113B970FD34F2CE 059B5E0851CF143A 86A560F10EC6D85B
 0170F175468FB5E6 0756D8E0774761D2 0CD3DA020021DC09
 43297FAD38E373FE 762514B829BF486A EA676B2CB7DB2B7A
 07A7137045DA2A16 3BDD119049372802 DFD64A815CAF1A0F
 04689104C2FD3B2F 26955F6835AF609A 5C513C9C4886C088
 37D06BB516CB7546 164D5E404F275232 0A2AEEAE3FF4AB77
 1F08260D1AC2465E 6B056E18759F5CCA EF1BF03E5DFA575A
 584023641ABA6176 004BD6EF09176062 88BF0DB6D70DEE56
 025816164629B007 480D39006EE762F2 A1F9915541020B56
 49793EBC79B3258F 437540C8698F3CFA 6FBF1CAFCFFD0556
 4FB05E1515AB73A7 072D43A077075292 2F22E49BAB7CA1AC
 49E95D6D4CA229BF 02FE55778117F12A 5A6B612CC26CCE4A
 018310DC409B26D6 1D9D5C5018F728C2 5F4C038ED12B2E41
 1C587F1C13924FEF 305532286D6F295A 63FAC0D034D9F793
*** End of Test ***
```

Chapter 8: Sorting

Chapter 8 of Numerical Recipes covers a variety of sorting tasks including sorting arrays into numerical order, preparing an index table for the order of an array, and preparing a rank table showing the rank order of each element in the array. PIKSRT *sorts a single array by the straight insertion method.* PIKSR2 *sorts by the same method but makes the corresponding rearrangement of a second array as well.* SHELL *carries out a Shell sort.* SORT *and* SORT2 *both do a Heapsort, and they are related in the same way as* PIKSRT *and* PIKSR2. *That is,* SORT *sorts a single array;* SORT2 *sorts an array while correspondingly rearranging a second array.* QCKSRT *sorts an array by the Quicksort algorithm, which is fast (on average) but requires a small amount of auxiliary storage.*

INDEXX *indexes an array. That is, it produces a second array that contains pointers to the elements of the original array in the order of their size.* SORT3 *uses* INDEXX *and illustrates its value by sorting one array while making corresponding rearrangements in two others.* RANK *produces the rank table for an array of data. The rank table is a second array whose elements list the rank order of the corresponding elements of the original array.*

Finally, the routines ECLASS *and* ECLAZZ *deal with equivalence classes.* ECLASS *gives the equivalence class of each element in an array based on a list of equivalent pairs which it is given as input.* ECLAZZ *gives the same output but bases it on a procedure named* EQUIV(J,K) *which tells whether two array elements* J *and* K *are in the same equivalence class.*

⋆ ⋆ ⋆ ⋆

Routine PIKSRT sorts an array by straight insertion. Sample program D8R1 provides it with a 100-element array from file TARRAY.DAT which is listed in the Appendix to this chapter. The program prints both the original and the sorted array for comparison.

```
      PROGRAM D8R1
C     Driver for routine PIKSRT
      DIMENSION A(100)
      OPEN(5,FILE='TARRAY.DAT',STATUS='OLD')
      READ(5,*) (A(I),I=1,100)
      CLOSE(5)
C     Print original array
      WRITE(*,*) 'Original array:'
      DO 11 I=1,10
        WRITE(*,'(1X,10F6.2)') (A(10*(I-1)+J),J=1,10)
11    CONTINUE
```

```
C       Sort array
        CALL PIKSRT(100,A)
C       Print sorted array
        WRITE(*,*) 'Sorted array:'
        DO 12 I=1,10
            WRITE(*,'(1X,10F6.2)') (A(10*(I-1)+J),J=1,10)
12      CONTINUE
        END
```

PIKSR2 sorts an array, and simultaneously rearranges a second array (of the same size) correspondingly. In program D8R2, the first array A(I) is again taken from TARRAY.DAT. The second is defined by B(I)=I. In other words, B is originally sorted and A is not. After a call to PIKSR2, the situation should be reversed. With a second call, this time with B as the first argument and A as the second, the two arrays should be returned to their original form.

```
        PROGRAM D8R2
C       Driver for routine PIKSR2
        DIMENSION A(100),B(100)
        OPEN(5,FILE='TARRAY.DAT',STATUS='OLD')
        READ(5,*) (A(I),I=1,100)
        CLOSE(5)
C       Generate B-array
        DO 11 I=1,100
            B(I)=I
11      CONTINUE
C       Sort A and mix B
        CALL PIKSR2(100,A,B)
        WRITE(*,*) 'After sorting A and mixing B, array A is:'
        DO 12 I=1,10
            WRITE(*,'(1X,10F6.2)') (A(10*(I-1)+J), J=1,10)
12      CONTINUE
        WRITE(*,*) '...and array B is:'
        DO 13 I=1,10
            WRITE(*,'(1X,10F6.2)') (B(10*(I-1)+J), J=1,10)
13      CONTINUE
        WRITE(*,*) 'press RETURN to continue...'
        READ(*,*)
C       Sort B and mix A
        CALL PIKSR2(100,B,A)
        WRITE(*,*) 'After sorting B and mixing A, array A is:'
        DO 14 I=1,10
            WRITE(*,'(1X,10F6.2)') (A(10*(I-1)+J), J=1,10)
14      CONTINUE
        WRITE(*,*) '...and array B is:'
        DO 15 I=1,10
            WRITE(*,'(1X,10F6.2)') (B(10*(I-1)+J), J=1,10)
15      CONTINUE
        END
```

Subroutine SHELL does a Shell sort of a data array. The calling format is identical to that of PIKSRT, and so we use the same sample program, now called D8R9.

```
        PROGRAM D8R9
C       Driver for routine SHELL
        DIMENSION A(100)
        OPEN(5,FILE='TARRAY.DAT',STATUS='OLD')
```

```
      READ(5,*) (A(I),I=1,100)
      CLOSE(5)
C     Print original array
      WRITE(*,*) 'Original array:'
      DO 11 I=1,10
          WRITE(*,'(1X,10F6.2)') (A(10*(I-1)+J),J=1,10)
11    CONTINUE
C     Sort array
      CALL SHELL(100,A)
C     Print sorted array
      WRITE(*,*) 'Sorted array:'
      DO 12 I=1,10
          WRITE(*,'(1X,10F6.2)') (A(10*(I-1)+J),J=1,10)
12    CONTINUE
      END
```

By the same token, routines SORT and SORT2 employ the same programs as routines PIKSRT and PIKSR2, respectively. (Here they are called D8R3 and D8R4.) Both routines use the Heapsort algorithm. SORT, however, works on a single array. SORT2 sorts one array while making corresponding rearrangements to a second.

```
      PROGRAM D8R3
C     Driver for routine SORT
      DIMENSION A(100)
      OPEN(5,FILE='TARRAY.DAT',STATUS='OLD')
      READ(5,*) (A(I),I=1,100)
      CLOSE(5)
C     Print original array
      WRITE(*,*) 'Original array:'
      DO 11 I=1,10
          WRITE(*,'(1X,10F6.2)') (A(10*(I-1)+J),J=1,10)
11    CONTINUE
C     Sort array
      CALL SORT(100,A)
C     Print sorted array
      WRITE(*,*) 'Sorted array:'
      DO 12 I=1,10
          WRITE(*,'(1X,10F6.2)') (A(10*(I-1)+J),J=1,10)
12    CONTINUE
      END

      PROGRAM D8R4
C     Driver for routine SORT2
      DIMENSION A(100),B(100)
      OPEN(5,FILE='TARRAY.DAT',STATUS='OLD')
      READ(5,*) (A(I),I=1,100)
      CLOSE(5)
C     Generate B-array
      DO 11 I=1,100
          B(I)=I
11    CONTINUE
C     Sort A and mix B
      CALL SORT2(100,A,B)
      WRITE(*,*) 'After sorting A and mixing B, array A is:'
      DO 12 I=1,10
          WRITE(*,'(1X,10F6.2)') (A(10*(I-1)+J), J=1,10)
12    CONTINUE
```

```
      WRITE(*,*) '...and array B is:'
      DO 13 I=1,10
         WRITE(*,'(1X,10F6.2)') (B(10*(I-1)+J), J=1,10)
13    CONTINUE
      WRITE(*,*) 'press RETURN to continue...'
      READ(*,*)
C     Sort B and mix A
      CALL SORT2(100,B,A)
      WRITE(*,*) 'After sorting B and mixing A, array A is:'
      DO 14 I=1,10
         WRITE(*,'(1X,10F6.2)') (A(10*(I-1)+J), J=1,10)
14    CONTINUE
      WRITE(*,*) '...and array B is:'
      DO 15 I=1,10
         WRITE(*,'(1X,10F6.2)') (B(10*(I-1)+J), J=1,10)
15    CONTINUE
      END
```

The subroutine INDEXX generates the index array for a given input array. The index array INDX(J) gives, for each J, the index of the element of the input array which will assume position J if the array is sorted. That is, for an input array A, the sorted version of A will be A(INDX(J)). To demonstrate this, sample program D8R5 produces an index for the array in TARRAY.DAT. It then prints the array in the order A(INDX(J)), J=1,...,100 for inspection.

```
      PROGRAM D8R5
C     Driver for routine INDEXX
      DIMENSION A(100),INDX(100)
      OPEN(5,FILE='TARRAY.DAT',STATUS='OLD')
      READ(5,*) (A(I),I=1,100)
      CLOSE(5)
C     Generate index for sorted array
      CALL INDEXX(100,A,INDX)
C     Print original array
      WRITE(*,*) 'Original array:'
      DO 11 I=1,10
         WRITE(*,'(1X,10F6.2)') (A(10*(I-1)+J),J=1,10)
11    CONTINUE
C     Print sorted array
      WRITE(*,*) 'Sorted array:'
      DO 12 I=1,10
         WRITE(*,'(1X,10F6.2)') (A(INDX(10*(I-1)+J)),J=1,10)
12    CONTINUE
      END
```

One use for INDEXX is the management of more than two arrays. SORT3, for example, sorts one array while making corresponding reorderings of two other arrays. In sample program D8R6, the first array is taken as the first 64 elements of TARRAY.DAT (see Appendix). The second and third arrays are taken to be the numbers 1 to 64 in forward order and reverse order, respectively. When the first array is ordered, the second and third are scrambled, but scrambled in exactly the same way. To prove this, a text message is assigned to a character array. Then the letters are scrambled according to the order of numbers found in the rearranged second array. They are subsequently unscrambled according to the order of numbers found in the rearranged third array. If SORT3 works properly, this ought to leave the message reading in the reverse order.

```
      PROGRAM D8R6
C     Driver for routine SORT3
      PARAMETER(NLEN=64)
      DIMENSION A(NLEN),IB(NLEN),IC(NLEN),WKSP(NLEN),INDX(NLEN)
      CHARACTER MSG1*33,MSG2*31
      CHARACTER MSG*64,AMSG(64)*1,BMSG(64)*1,CMSG(64)*1
      EQUIVALENCE(MSG,AMSG(1)),(MSG1,AMSG(1)),(MSG2,AMSG(34))
      DATA MSG1/'I''d rather have a bottle in front'/
      DATA MSG2/' of me than a frontal lobotomy.'/
      WRITE(*,*) 'Original message:'
      WRITE(*,'(1X,64A1,/)') (AMSG(J),J=1,64)
C     Read array of random numbers
      OPEN(5,FILE='TARRAY.DAT',STATUS='OLD')
      READ(5,*) (A(I),I=1,NLEN)
      CLOSE(5)
C     Create array IB and array IC
      DO 11 I=1,NLEN
        IB(I)=I
        IC(I)=NLEN+1-I
11    CONTINUE
C     Sort array A while mixing IB and IC
      CALL SORT3(NLEN,A,IB,IC,WKSP,INDX)
C     Scramble message according to array IB
      DO 12 I=1,NLEN
        J=IB(I)
        BMSG(I)=AMSG(J)
12    CONTINUE
      WRITE(*,*) 'Scrambled message:'
      WRITE(*,'(1X,64A1,/)') (BMSG(J),J=1,64)
C     Unscramble according to array C
      DO 13 I=1,NLEN
        J=IC(I)
        CMSG(J)=BMSG(I)
13    CONTINUE
      WRITE(*,*) 'Mirrored message:'
      WRITE(*,'(1X,64A1,/)') (CMSG(J),J=1,64)
      END
```

RANK is a subroutine that is similar to INDEXX. Instead of producing an indexing array, though, it produces a rank table. For an array A(J) and rank table IRANK(J), entry J in IRANK will tell what index A(J) will have if A is sorted. IRANK actually takes its input information not from the array itself, but from the index array produced by INDEXX. Sample program D8R7 begins with the array from TARRAY, and feeds it to INDEXX and RANK. The table of ranks produced is listed. To check it, the array A is copied into an array B in the rank order suggested by IRANK. B should then be in proper order.

```
      PROGRAM D8R7
C     Driver for routine RANK
      DIMENSION A(100),B(10),INDX(100),IRANK(100)
      OPEN(5,FILE='TARRAY.DAT',STATUS='OLD')
      READ(5,*) (A(I),I=1,100)
      CLOSE(5)
      CALL INDEXX(100,A,INDX)
      CALL RANK(100,INDX,IRANK)
```

```
      WRITE(*,*) 'Original array is:'
      DO 11 I=1,10
        WRITE(*,'(1X,10F6.2)') (A(10*(I-1)+J), J=1,10)
11    CONTINUE
      WRITE(*,*) 'Table of ranks is:'
      DO 12 I=1,10
        WRITE(*,'(1X,10I6)') (IRANK(10*(I-1)+J), J=1,10)
12    CONTINUE
      WRITE(*,*) 'press RETURN to continue...'
      READ(*,*)
      WRITE(*,*) 'Array sorted according to rank table:'
      DO 15 I=1,10
        DO 14 J=1,10
          K=10*(I-1)+J
          DO 13 L=1,100
            IF (IRANK(L).EQ.K) B(J)=A(L)
13        CONTINUE
14      CONTINUE
        WRITE(*,'(1X,10F6.2)') (B(J),J=1,10)
15    CONTINUE
      END
```

QCKSRT sorts an array by the Quicksort algorithm. Its calling sequence is exactly like that of PIKSRT and SORT, so we again rely on the same sample program, now called D8R8.

```
      PROGRAM D8R8
C     Driver for routine QCKSRT
      DIMENSION A(100)
      OPEN(5,FILE='TARRAY.DAT',STATUS='OLD')
      READ(5,*) (A(I),I=1,100)
      CLOSE(5)
C     Print original array
      WRITE(*,*) 'Original array:'
      DO 11 I=1,10
        WRITE(*,'(1X,10F6.2)') (A(10*(I-1)+J),J=1,10)
11    CONTINUE
C     Sort array
      CALL QCKSRT(100,A)
C     Print sorted array
      WRITE(*,*) 'Sorted array:'
      DO 12 I=1,10
        WRITE(*,'(1X,10F6.2)') (A(10*(I-1)+J),J=1,10)
12    CONTINUE
      END
```

Subroutine ECLASS generates a list of equivalence classes for the elements of an input array, based on the arrays LISTA(J) and LISTB(J) which list equivalent pairs for each J. In sample program D8R10, these lists are

$$\begin{array}{ll} \text{LISTA}: & 1,1,5,2,6,2,7,11,3,4,12 \\ \text{LISTB}: & 5,9,13,6,10,14,3,7,15,8,4 \end{array}$$

According to these lists, 1 is equivalent to 5, 1 is equivalent to 9, etc. If you work it out, you will find the following classes:

$$\begin{array}{ll} \texttt{class1}: & 1,5,9,13 \\ \texttt{class2}: & 2,6,10,14 \\ \texttt{class3}: & 3,7,11,15 \\ \texttt{class4}: & 4,8,12 \end{array}$$

The sample program prints out the classes and ought to agree with this list.

```
      PROGRAM D8R10
C     Driver for routine ECLASS
      PARAMETER(N=15,M=11)
      DIMENSION LISTA(M),LISTB(M),NF(N),NFLAG(N),NSAV(N)
      DATA LISTA/1,1,5,2,6,2,7,11,3,4,12/
      DATA LISTB/5,9,13,6,10,14,3,7,15,8,4/
      CALL ECLASS(NF,N,LISTA,LISTB,M)
      DO 11 I=1,N
        NFLAG(I)=1
11    CONTINUE
      WRITE(*,'(/1X,A)') 'Numbers from 1-15 divided according to'
      WRITE(*,'(1X,A/)') 'their value modulo 4:'
      LCLAS=0
      DO 13 I=1,N
        NCLASS=NF(I)
        IF(NFLAG(NCLASS).NE.0) THEN
          NFLAG(NCLASS)=0
          LCLAS=LCLAS+1
          K=0
          DO 12 J=I,N
            IF (NF(J).EQ.NF(I)) THEN
              K=K+1
              NSAV(K)=J
            ENDIF
12        CONTINUE
          WRITE(*,'(1X,A,I2,A,3X,5I3)') 'Class',
     *          LCLAS,':',(NSAV(J),J=1,K)
        ENDIF
13    CONTINUE
      END
```

ECLAZZ performs the same analysis but figures the equivalences from a logical function `EQUIV(I,J)` that tells whether I and J are in the same equivalence class. In D8R11, EQUIV is defined as `.TRUE.` if `(I MOD 4)` and `(J MOD 4)` are the same. It is otherwise `.FALSE.`

```
      PROGRAM D8R11
C     Driver for routine ECLAZZ
      EXTERNAL EQUIV
      LOGICAL EQUIV
      PARAMETER(N=15)
      DIMENSION NF(N),NFLAG(N),NSAV(N)
      CALL ECLAZZ(NF,N,EQUIV)
      DO 11 I=1,N
        NFLAG(I)=1
11    CONTINUE
      WRITE(*,'(/1X,A)') 'Numbers from 1-15 divided according to'
      WRITE(*,'(1X,A/)') 'their value modulo 4:'
```

```
      LCLAS=0
      DO 13 I=1,N
          NCLASS=NF(I)
          IF(NFLAG(NCLASS).NE.0) THEN
              NFLAG(NCLASS)=0
              LCLAS=LCLAS+1
              K=0
              DO 12 J=I,N
                  IF (NF(J).EQ.NF(I)) THEN
                      K=K+1
                      NSAV(K)=J
                  ENDIF
12            CONTINUE
              WRITE(*,'(1X,A,I2,A,3X,5I3)') 'Class',
     *                LCLAS,':',(NSAV(J),J=1,K)
          ENDIF
13    CONTINUE
      END
      LOGICAL FUNCTION EQUIV(I,J)
      EQUIV=.FALSE.
      IF (MOD(I,4).EQ.MOD(J,4)) EQUIV=.TRUE.
      RETURN
      END
```

Appendix

File TARRAY.DAT:

```
29.82 71.51  3.30 87.44 53.42 63.16 89.10 25.75 93.16 27.72
71.58 48.34 53.11 18.34 27.13 60.31 83.34 22.81 66.84 52.91
53.42 15.22  8.01 53.39 76.12 79.09 67.61 38.39 24.81 73.21
13.42 52.10 34.86 99.83 38.46 81.59 61.75 79.62 93.39  3.21
99.34 92.22 94.29  7.03  6.67 89.35 83.14  9.01 12.68 62.22
 2.95 85.02 95.82 73.96 49.29 77.72 36.65  3.48 48.98 71.83
 1.41  9.48 32.37 89.95 28.39 79.36 54.05 46.08 11.67 37.78
77.17 74.33 10.13  4.62 49.95 68.40 19.40 34.06  4.11 98.40
42.44 64.14 89.41 52.99 71.79  3.94 19.73 44.91 71.44 59.10
27.54 15.67 67.95 55.61 26.05 25.01 82.09 89.67 57.08 38.27
```

Chapter 9: Root Finding and Sets of Equations

Chapter 9 of Numerical Recipes deals primarily with the problem of finding roots to equations, and treats the problem in greatest detail in one dimension. We begin with a general-purpose routine called SCRSHO *that produces a crude graph of a given function on a specified interval. It is used for low-resolution plotting to investigate the properties of the function. With this in hand, we add bracketing routines* ZBRAC *and* ZBRAK*. The first of these takes a function and an interval, and expands the interval geometrically until it brackets a root. The second breaks the interval into N subintervals of equal size. It then reports any intervals that contain at least one root. Once bracketed, roots can be found by a number of other routines.* RTBIS *finds such roots by bisection.* RTFLSP *and* RTSEC *use the method of false position and the secant method, respectively.* ZBRENT *uses a combination of methods to give assured and relatively efficient convergence.* RTNEWT *implements the Newton-Raphson root finding method, while* RTSAFE *combines it with bisection to correct for its risky global convergence properties.*

For finding the roots of polynomials, LAGUER *is handy, and when combined with its driver* ZROOTS *it can find all roots of a polynomial having complex coefficients. When you have some tentative complex roots of a real polynomial, they can be polished by* QROOT*, which employs Bairstow's method.*

In multiple dimensions, root-finding requires some foresight. However, if you can identify the neighborhood of a root of a system of nonlinear equations, then MNEWT *will help you to zero in using Newton-Raphson.*

⋆ ⋆ ⋆ ⋆

SCRSHO is a primitive graphing routine that will print graphs on virtually any terminal or printer. Sample program D9R1 demonstrates it by graphing the zero-order Bessel function J_0.

```
      PROGRAM D9R1
C     Driver for routine SCRSHO
      EXTERNAL BESSJ0
      WRITE(*,*) 'Graph of the Bessel Function J0:'
      CALL SCRSHO(BESSJ0)
      END
```

ZBRAC is a root-bracketing routine that works by expanding the range of an interval geometrically until it brackets a root. Sample program D9R2 applies it to the Bessel function J_0. It starts with the ten intervals (1.0, 2.0), (2.0, 3.0), etc., and expands each until it contains a root. Then it prints the interval limits, and the function J_0 evaluated at these limits. The two values of J_0 should have opposite signs.

```
      PROGRAM D9R2
C     Driver for routine ZBRAC
      LOGICAL SUCCES
      EXTERNAL BESSJO
      WRITE(*,'(/1X,T4,A,T29,A/))') 'Bracketing values:',
     *         'Function values:'
      WRITE(*,'(1X,T6,A,T16,A,T29,A,T41,A/))') 'X1','X2',
     *         'BESSJO(X1)','BESSJO(X2)'
      DO 11 I=1,10
        X1=I
        X2=X1+1.0
        CALL ZBRAC(BESSJO,X1,X2,SUCCES)
        IF (SUCCES) THEN
          WRITE(*,'(1X,F7.2,F10.2,7X,2F12.6))') X1,X2,
     *              BESSJO(X1),BESSJO(X2)
        ENDIF
11    CONTINUE
      END
```

ZBRAK is much like ZBRAC except that it takes an interval and subdivides it into N equal parts. It then identifies any of the subintervals that contain roots. Sample program D9R3 looks for roots of $J_0(x)$ between X1 $= 1.0$ and X2 $= 50.0$ by allowing ZBRAK to divide the interval into $N = 100$ parts. If there are no roots spaced closer than $\Delta x = 0.49$, then it will find brackets for all roots in this region. The limits of bracketing intervals, as well as function values at these limits, are printed, and again the function values at the end of each interval ought to be of opposite sign. There are 16 roots of J_0 between 1 and 50.

```
      PROGRAM D9R3
C     Driver for routine ZBRAK
      EXTERNAL BESSJO
      PARAMETER(N=100,NBMAX=20,X1=1.0,X2=50.0)
      DIMENSION XB1(NBMAX),XB2(NBMAX)
      NB=NBMAX
      CALL ZBRAK(BESSJO,X1,X2,N,XB1,XB2,NB)
      WRITE(*,'(/1X,A/)') 'Brackets for roots of BESSJO:'
      WRITE(*,'(/1X,T17,A,T27,A,T40,A,T50,A/)') 'lower','upper',
     *         'F(lower)','F(upper)'
      DO 11 I=1,NB
        WRITE(*,'(1X,A,I2,2(4X,2F10.4))') 'Root ',I,XB1(I),XB2(I),
     *         BESSJO(XB1(I)),BESSJO(XB2(I))
11    CONTINUE
      END
```

Routine RTBIS begins with the brackets for a root and finds the root itself by bisection, The accuracy with which the root is found is determined by parameter XACC. Sample program D9R4 finds all the roots of Bessel function $J_0(x)$ between X1 $= 1.0$ and X2 $= 50.0$. In this case XACC is specified to be about 10^{-6} of the value of the root itself (actually, 10^{-6} of the center of the interval being bisected). The roots ROOT are listed, as well as J_0(ROOT) to verify their accuracy.

```
      PROGRAM D9R4
C     Driver for routine RTBIS
      EXTERNAL BESSJ0
      PARAMETER(N=100,NBMAX=20,X1=1.0,X2=50.0)
      DIMENSION XB1(NBMAX),XB2(NBMAX)
      NB=NBMAX
      CALL ZBRAK(BESSJ0,X1,X2,N,XB1,XB2,NB)
      WRITE(*,'(/1X,A)') 'Roots of BESSJ0:'
      WRITE(*,'(/1X,T19,A,T31,A/)') 'x','F(x)'
      DO 11 I=1,NB
        XACC=(1.0E-6)*(XB1(I)+XB2(I))/2.0
        ROOT=RTBIS(BESSJ0,XB1(I),XB2(I),XACC)
        WRITE(*,'(1X,A,I2,2X,F12.6,E16.4)') 'Root ',I,ROOT,BESSJ0(ROOT)
11    CONTINUE
      END
```

The next five sample programs, D9R5–D9R9, are essentially identical to the one just discussed, except for the root-finder they employ. D9R5 calls RTFLSP, finding the root by "false position". D9R6 calls RTSEC and uses the secant method. D9R7 uses ZBRENT to give reliable and efficient convergence. The Newton-Raphson method implemented in RTNEWT is demonstrated by D9R8, and D9R9 calls RTSAFE, which improves upon RTNEWT by combining it with bisection to achieve better global convergence. The latter two programs include a subroutine FUNCD that returns the value of the function and its derivative at a given x. In the case of test function $J_0(x)$ the derivative is $-J_1(x)$, and is conveniently in our collection of special functions.

```
      PROGRAM D9R5
C     Driver for routine RTFLSP
      EXTERNAL BESSJ0
      PARAMETER(N=100,NBMAX=20,X1=1.0,X2=50.0)
      DIMENSION XB1(NBMAX),XB2(NBMAX)
      NB=NBMAX
      CALL ZBRAK(BESSJ0,X1,X2,N,XB1,XB2,NB)
      WRITE(*,'(/1X,A)') 'Roots of BESSJ0:'
      WRITE(*,'(/1X,T19,A,T31,A/)') 'x','F(x)'
      DO 11 I=1,NB
        XACC=(1.0E-6)*(XB1(I)+XB2(I))/2.0
        ROOT=RTFLSP(BESSJ0,XB1(I),XB2(I),XACC)
        WRITE(*,'(1X,A,I2,2X,F12.6,E16.4)') 'Root ',I,ROOT,BESSJ0(ROOT)
11    CONTINUE
      END
```

```
      PROGRAM D9R6
C     Driver for routine RTSEC
      EXTERNAL BESSJ0
      PARAMETER(N=100,NBMAX=20,X1=1.0,X2=50.0)
      DIMENSION XB1(NBMAX),XB2(NBMAX)
      NB=NBMAX
      CALL ZBRAK(BESSJ0,X1,X2,N,XB1,XB2,NB)
      WRITE(*,'(/1X,A)') 'Roots of BESSJ0:'
      WRITE(*,'(/1X,T19,A,T31,A/)') 'x','F(x)'
      DO 11 I=1,NB
        XACC=(1.0E-6)*(XB1(I)+XB2(I))/2.0
        ROOT=RTSEC(BESSJ0,XB1(I),XB2(I),XACC)
        WRITE(*,'(1X,A,I2,2X,F12.6,E16.4)') 'Root ',I,ROOT,BESSJ0(ROOT)
11    CONTINUE
```

```
      END

      PROGRAM D9R7
C     Driver for routine ZBRENT
      EXTERNAL BESSJ0
      PARAMETER(N=100,NBMAX=20,X1=1.0,X2=50.0)
      DIMENSION XB1(NBMAX),XB2(NBMAX)
      NB=NBMAX
      CALL ZBRAK(BESSJ0,X1,X2,N,XB1,XB2,NB)
      WRITE(*,'(/1X,A)') 'Roots of BESSJ0:'
      WRITE(*,'(/1X,T19,A,T31,A/)') 'x','F(x)'
      DO 11 I=1,NB
        TOL=(1.0E-6)*(XB1(I)+XB2(I))/2.0
        ROOT=ZBRENT(BESSJ0,XB1(I),XB2(I),TOL)
        WRITE(*,'(1X,A,I2,2X,F12.6,E16.4)') 'Root ',I,ROOT,BESSJ0(ROOT)
11    CONTINUE
      END

      PROGRAM D9R8
C     Driver for routine RTNEWT
      EXTERNAL FUNCD,BESSJ0
      PARAMETER(N=100,NBMAX=20,X1=1.0,X2=50.0)
      DIMENSION XB1(NBMAX),XB2(NBMAX)
      NB=NBMAX
      CALL ZBRAK(BESSJ0,X1,X2,N,XB1,XB2,NB)
      WRITE(*,'(/1X,A)') 'Roots of BESSJ0:'
      WRITE(*,'(/1X,T19,A,T31,A/)') 'x','F(x)'
      DO 11 I=1,NB
        XACC=(1.0E-6)*(XB1(I)+XB2(I))/2.0
        ROOT=RTNEWT(FUNCD,XB1(I),XB2(I),XACC)
        WRITE(*,'(1X,A,I2,2X,F12.6,E16.4)') 'Root ',I,ROOT,BESSJ0(ROOT)
11    CONTINUE
      END
      SUBROUTINE FUNCD(X,FN,DF)
      FN=BESSJ0(X)
      DF=-BESSJ1(X)
      RETURN
      END

      PROGRAM D9R9
C     Driver for routine RTSAFE
      EXTERNAL FUNCD,BESSJ0
      PARAMETER(N=100,NBMAX=20,X1=1.0,X2=50.0)
      DIMENSION XB1(NBMAX),XB2(NBMAX)
      NB=NBMAX
      CALL ZBRAK(BESSJ0,X1,X2,N,XB1,XB2,NB)
      WRITE(*,'(/1X,A)') 'Roots of BESSJ0:'
      WRITE(*,'(/1X,T19,A,T31,A/)') 'x','F(x)'
      DO 11 I=1,NB
        XACC=(1.0E-6)*(XB1(I)+XB2(I))/2.0
        ROOT=RTSAFE(FUNCD,XB1(I),XB2(I),XACC)
        WRITE(*,'(1X,A,I2,2X,F12.6,E16.4)') 'Root ',I,ROOT,BESSJ0(ROOT)
11    CONTINUE
      END
      SUBROUTINE FUNCD(X,FN,DF)
      FN=BESSJ0(X)
      DF=-BESSJ1(X)
```

```
      RETURN
      END
```

Routine LAGUER finds the roots of a polynomial with complex coefficients. The polynomial of degree M is specified by $M+1$ coefficients which, in sample program D9R10, are specified in a DATA statement and kept in array A. The polynomial in this case is

$$F(x) = x^4 - (1+2i)x^2 + 2i$$

The four roots of this polynomial are $x = 1.0$, $x = -1.0$, $x = 1+i$, and $x = -(1+i)$. LAGUER proceeds on the basis of a trial root, and attempts to converge to true roots. The root to which it converges depends on the trial value. The program tries a series of complex trial values along the line in the imaginary plane from $-1.0-i$ to $1.0+i$. The actual roots to which it converges are compared to all previously found values, and if different, are printed.

```
      PROGRAM D9R10
C     Driver for routine LAGUER
      PARAMETER(M=4,MP1=M+1,NTRY=21,EPS=1.0E-6)
      COMPLEX A(MP1),Y(NTRY),X
      LOGICAL POLISH
      DATA A/(0.0,2.0),(0.0,0.0),(-1.0,-2.0),(0.0,0.0),(1.0,0.0)/
      WRITE(*,'(/1X,A)') 'Roots of polynomial x^4-(1+2i)*x^2+2i'
      WRITE(*,'(/1X,T16,A,T29,A/)') 'Real','Complex'
      N=0
      POLISH=.FALSE.
      DO 12 I=1,NTRY
        X=CMPLX((I-11.0)/10.0,(I-11.0)/10.0)
        CALL LAGUER(A,M,X,EPS,POLISH)
        IF (N.EQ.0) THEN
          N=1
          Y(1)=X
          WRITE(*,'(1X,I5,2F15.6)') N,X
        ELSE
          IFLAG=0
          DO 11 J=1,N
            IF (CABS(X-Y(J)).LE.
     *              EPS*CABS(X)) IFLAG=1
11        CONTINUE
          IF (IFLAG.EQ.0) THEN
            N=N+1
            Y(N)=X
            WRITE(*,'(1X,I5,2F15.6)') N,X
          ENDIF
        ENDIF
12    CONTINUE
      END
```

ZROOTS is a driver for LAGUER. Sample program D9R11 exercises ZROOTS, using the same polynomial as the previous routine. First it finds the four roots. Then it corrupts each one by multiplying by 1.01. Finally it uses ZROOTS again to polish the corrupted roots by setting the logical parameter POLISH to .TRUE.

```
      PROGRAM D9R11
C     Driver for routine ZROOTS
      PARAMETER(M=4,M1=M+1)
      COMPLEX A(M1),X,ROOTS(M)
      LOGICAL POLISH
      DATA A/(0.0,2.0),(0.0,0.0),(-1.0,-2.0),(0.0,0.0),(1.0,0.0)/
      WRITE(*,'(/1X,A)') 'Roots of the polynomial x^4-(1+2i)*x^2+2i'
      POLISH=.FALSE.
      CALL ZROOTS(A,M,ROOTS,POLISH)
      WRITE(*,'(/1X,A)') 'Unpolished roots:'
      WRITE(*,'(1X,T10,A,T25,A,T37,A)') 'Root #','Real','Imag.'
      DO 11 I=1,M
        WRITE(*,'(1X,I11,5X,2F12.6)') I,ROOTS(I)
11    CONTINUE
      WRITE(*,'(/1X,A)') 'Corrupted roots:'
      DO 12 I=1,M
        ROOTS(I)=ROOTS(I)*(1.0+0.01*I)
12    CONTINUE
      WRITE(*,'(1X,T10,A,T25,A,T37,A)') 'Root #','Real','Imag.'
      DO 13 I=1,M
        WRITE(*,'(1X,I11,5X,2F12.6)') I,ROOTS(I)
13    CONTINUE
      POLISH=.TRUE.
      CALL ZROOTS(A,M,ROOTS,POLISH)
      WRITE(*,'(/1X,A)') 'Polished roots:'
      WRITE(*,'(1X,T10,A,T25,A,T37,A)') 'Root #','Real','Imag.'
      DO 14 I=1,M
        WRITE(*,'(1X,I11,5X,2F12.6)') I,ROOTS(I)
14    CONTINUE
      END
```

QROOT is used for finding quadratic factors of polynomials with real coefficients. In the case of sample program D9R12, the polynomial is

$$P(x) = x^6 - 6x^5 + 16x^4 - 24x^3 + 25x^2 - 18x + 10.$$

The program proceeds like that of LAGUER. Successive trial values for quadratic factors $x^2 + Bx + C$ (in the form of guesses for B and C) are made, and for each trial, QROOT converges on correct values. If the B and C which are found are unlike any previous values, then they are printed. By this means, all three quadratic factors are located. You can, of course, compare their product to the polynomial above.

```
      PROGRAM D9R12
C     Driver for routine QROOT
      PARAMETER(N=7,EPS=1.0E-6,NTRY=10,TINY=1.0E-5)
      DIMENSION P(N),B(NTRY),C(NTRY)
      DATA P/10.0,-18.0,25.0,-24.0,16.0,-6.0,1.0/
      WRITE(*,'(/1X,A)') 'P(x)=x^6-6x^5+16x^4-24x^3+25x^2-18x+10'
      WRITE(*,'(1X,A)') 'Quadratic factors x^2+Bx+C'
      WRITE(*,'(/1X,A,T15,A,T27,A/)') 'Factor','B','C'
      NROOT=0
      DO 12 I=1,NTRY
        C(I)=0.5*I
        B(I)=-0.5*I
        CALL QROOT(P,N,B(I),C(I),EPS)
        IF (NROOT.EQ.0) THEN
```

```
                WRITE(*,'(1X,I3,2X,2F12.6)') NROOT,B(I),C(I)
                NROOT=1
            ELSE
                NFLAG=0
                DO 11 J=1,NROOT
                    IF (ABS(B(I)-B(J)).LT.TINY
     *                  .AND. ABS(C(I)-C(J)).LT.TINY) NFLAG=1
11              CONTINUE
                IF (NFLAG.EQ.0) THEN
                    WRITE(*,'(1X,I3,2X,2F12.6)') NROOT,B(I),C(I)
                    NROOT=NROOT+1
                ENDIF
            ENDIF
12      CONTINUE
        END
```

Finally, MNEWT looks for roots of multiple nonlinear equations. In order to run a sample program D9R13 we supply a subroutine USRFUN that returns the matrix ALPHA of partial derivatives of the functions with respect to each of the variables, and vector BETA, containing the negatives of the function values. The sample program tries to find sets of variables that solve the four equations

$$
\begin{aligned}
-x_1^2 - x_2^2 - x_3^2 + x_4 &= 0 \\
x_1^2 + x_2^2 + x_3^2 + x_4^2 - 1 &= 0 \\
x_1 - x_2 &= 0 \\
x_2 - x_3 &= 0
\end{aligned}
$$

You will probably be able to find the two solutions to this set even without MNEWT, noting that $x_1 = x_2$ and $x_2 = x_3$. If not, simply take the output from MNEWT and plug it into these equations for verification. The output from MNEWT should convince you of the need for good starting values.

```
      PROGRAM D9R13
C     Driver for routine MNEWT
      PARAMETER(NTRIAL=5,TOLX=1.0E-6,N=4,TOLF=1.0E-6,NP=15)
      DIMENSION X(NP),ALPHA(NP,NP),BETA(NP)
      DO 15 KK=-1,1,2
          DO 14 K=1,3
              XX=0.2*K*KK
              WRITE(*,'(/1X,A,I2)') 'Starting vector number',K
              DO 11 I=1,4
                  X(I)=XX+0.2*I
                  WRITE(*,'(1X,T5,A,I1,A,F5.2)')'X(',I,') = ',X(I)
11            CONTINUE
              DO 13 J=1,NTRIAL
                  CALL MNEWT(1,X,N,TOLX,TOLF)
                  CALL USRFUN(X,ALPHA,BETA)
                  WRITE(*,'(/1X,T5,A,T14,A,T29,A/)')'I','X(I)','F'
                  DO 12 I=1,N
                      WRITE(*,'(1X,I4,2E15.6)')I,X(I),-BETA(I)
12                CONTINUE
                  WRITE(*,'(/1X,A)') 'press RETURN to continue...'
                  READ(*,*)
13            CONTINUE
14        CONTINUE
```

```
15    CONTINUE
      END
      SUBROUTINE USRFUN(X,ALPHA,BETA)
      PARAMETER(NP=15)
      DIMENSION ALPHA(NP,NP),BETA(NP),X(NP)
      ALPHA(1,1)=-2.0*X(1)
      ALPHA(1,2)=-2.0*X(2)
      ALPHA(1,3)=-2.0*X(3)
      ALPHA(1,4)=1.0
      ALPHA(2,1)=2.0*X(1)
      ALPHA(2,2)=2.0*X(2)
      ALPHA(2,3)=2.0*X(3)
      ALPHA(2,4)=2.0*X(4)
      ALPHA(3,1)=1.0
      ALPHA(3,2)=-1.0
      ALPHA(3,3)=0.0
      ALPHA(3,4)=0.0
      ALPHA(4,1)=0.0
      ALPHA(4,2)=1.0
      ALPHA(4,3)=-1.0
      ALPHA(4,4)=0.0
      BETA(1)=X(1)**2+X(2)**2+X(3)**2-X(4)
      BETA(2)=-X(1)**2-X(2)**2-X(3)**2-X(4)**2+1.0
      BETA(3)=-X(1)+X(2)
      BETA(4)=-X(2)+X(3)
      END
```

Chapter 10: Minimization and Maximization of Functions

Chapter 10 of Numerical Recipes deals with finding the maxima and minima of functions. The task has two parts, first the discovery of one or more bracketing intervals, and then the convergence to an extremum. MNBRAK *begins with two specified abscissas of a function and searches in the "downhill" direction for brackets of a minimum.* GOLDEN *can then take a bracketing triplet and perform a golden section search to a specified precision, for the minimum itself. When you are not concerned with worst-case examples, but only very efficient average-case performance, Brent's method (routine* BRENT*) is recommended. In the event that means are at hand for calculating the function's derivative as well as its value, consider* DBRENT.

Multidimensional minimization strategies may be based on the one-dimensional algorithms. Our single example of an algorithm that is not so based is AMOEBA*, which utilizes the downhill simplex method. Among the ones that do use one-dimensional methods are* POWELL, FRPRMN, *and* DFPMIN. *These three all make calls to* LINMIN*, a subroutine that minimizes a function along a given direction in space.* LINMIN *in turn uses the one-dimensional algorithm* BRENT*, if derivatives are not known, or* DBRENT *if they are.* POWELL *uses only function values and minimizes along an artfully chosen set of favorable directions.* FRPRMN *uses a Fletcher-Reeves-Polak-Ribiere minimization and requires the calculation of derivatives for the function.* DFPMIN *uses a variant of the Davidon-Fletcher-Powell variable metric method. This, too, requires calculation of derivatives.*

The chapter ends with two topics of somewhat different nature. The first is linear programming, which deals with the maximation of a linear combination of variables, subject to linear constraints. This problem is dealt with by the simplex method in routine SIMPLX. *The second is the subject of large scale optimization, which is illustrated with the method of simulated annealing, and applied particularly to the "travelling salesman" problem in routine* ANNEAL.

★ ★ ★ ★

MNBRAK searches a given function for a minimum. Given two values AX and BX of abscissa, it searches in the downward direction until it can find three new values AX,BX,CX that bracket a minimum. FA,FB,FC are the values of the function at these points. Sample program D10R1 is a simple application of MNBRAK applied to the Bessel function J_0. It tries a series of starting values AX,BX each encompassing an interval of length 1.0. MNBRAK then finds several bracketing intervals of various minima of J_0.

```
      PROGRAM D10R1
C     Driver for routine MNBRAK
      EXTERNAL BESSJ0
      DO 11 I=1,10
        AX=I*0.5
        BX=(I+1.0)*0.5
        CALL MNBRAK(AX,BX,CX,FA,FB,FC,BESSJ0)
        WRITE(*,'(1X,T13,A,T25,A,T37,A)') 'A','B','C'
        WRITE(*,'(1X,A3,T5,3F12.6)') 'X',AX,BX,CX
        WRITE(*,'(1X,A3,T5,3F12.6)') 'F',FA,FB,FC
11    CONTINUE
      END
```

Routine GOLDEN continues the minimization process by taking a bracketing triplet AX,BX,CX and performing a golden section search to isolate the contained minimum to a stated precision TOL. Sample program D10R2 again uses J_0 as the test function. Using intervals (AX,BX) of length 1.0 it uses MNBRAK to bracket all minima between $x = 0.0$ and $x = 100.0$. Some minima are bracketed more than once. On each pass, the bracketed solution is tracked down by GOLDEN. It is then compared to all previously located minima, and if different it is added to the collection by incrementing NMIN (number of minima found) and adding the location XMIN of the minima to the list in array AMIN. As a check of GOLDEN, the routine prints out the value of J_0 at the minimum, and also the value of J_1, which ought to be zero at extrema of J_0.

```
      PROGRAM D10R2
C     Driver for routine GOLDEN
      PARAMETER(TOL=1.0E-6,EQL=1.E-3)
      EXTERNAL BESSJ0
      DIMENSION AMIN(20)
      NMIN=0
      WRITE(*,'(/1X,A)') 'Minima of the function BESSJ0'
      WRITE(*,'(/1X,T6,A,T19,A,T27,A,T40,A/)') 'Min. #','X',
     *       'BESSJ0(X)','BESSJ1(X)'
      DO 12 I=1,100
        AX=I
        BX=I+1.0
        CALL MNBRAK(AX,BX,CX,FA,FB,FC,BESSJ0)
        G=GOLDEN(AX,BX,CX,BESSJ0,TOL,XMIN)
        IF (NMIN.EQ.0) THEN
          AMIN(1)=XMIN
          NMIN=1
          WRITE(*,'(1X,5X,I2,3X,3F12.6)') NMIN,XMIN,
     *           BESSJ0(XMIN),BESSJ1(XMIN)
        ELSE
          IFLAG=0
          DO 11 J=1,NMIN
            IF (ABS(XMIN-AMIN(J)).LE.EQL*XMIN)
     *             IFLAG=1
11        CONTINUE
          IF (IFLAG.EQ.0) THEN
            NMIN=NMIN+1
            AMIN(NMIN)=XMIN
```

```
                WRITE(*,'(1X,5X,I2,3X,3F12.6)') NMIN,
     *                   XMIN,BESSJO(XMIN),BESSJ1(XMIN)
            ENDIF
          ENDIF
12      CONTINUE
        END
```

There are two other routines presented which also take the bracketing triplet AX,BX,CX from MNBRAK and find the contained minimum. They are BRENT and DBRENT. The sample programs for these two, D1OR3 and D1OR4, are virtually identical to that used on GOLDEN. Consequently, we can focus on the minor differences. For one thing, the driver for GOLDEN just passes the function name BESSJO to GOLDEN as an argument, and declares BESSJO as EXTERNAL to make this possible. In the program D1OR3 of BRENT, the function is called FUNC, and FUNC is declared EXTERNAL. Then a trivial FUNCTION subroutine is provided with FUNC(X) set equal to BESSJO(X). This illustrates two methods of accomplishing the same goal. The advantage of the second method is that if you are planning to deal with several different functions, you can avoid multiple compilations of the program; simply recompile the very short function subroutines and link each in turn to the same program. In program D1OR4 we combined methods by passing BESSJO as an argument, and defining its derivative (−BESSJ1) in an external function routine DERIV. Note that DBRENT is only used when the derivative can be calculated conveniently.

```
        PROGRAM D1OR3
C       Driver for routine BRENT
        PARAMETER(TOL=1.OE-6,EQL=1.E-4)
        DIMENSION AMIN(20)
        EXTERNAL FUNC
        NMIN=0
        WRITE(*,'(/1X,A)') 'Minima of the function BESSJO'
        WRITE(*,'(/1X,T6,A,T19,A,T28,A,T40,A/)') 'Min. #','X',
     *         'BESSJO(X)','BESSJ1(X)'
        DO 12 I=1,100
          AX=I
          BX=I+1.0
          CALL MNBRAK(AX,BX,CX,FA,FB,FC,FUNC)
          B=BRENT(AX,BX,CX,FUNC,TOL,XMIN)
          IF (NMIN.EQ.0) THEN
            AMIN(1)=XMIN
            NMIN=1
            WRITE(*,'(1X,5X,I2,3X,3F12.6)') NMIN,XMIN,
     *               BESSJO(XMIN),BESSJ1(XMIN)
          ELSE
            IFLAG=0
            DO 11 J=1,NMIN
              IF (ABS(XMIN-AMIN(J)).LE.EQL*XMIN)
     *                IFLAG=1
11          CONTINUE
            IF (IFLAG.EQ.0) THEN
              NMIN=NMIN+1
              AMIN(NMIN)=XMIN
              WRITE(*,'(1X,5X,I2,3X,3F12.6)') NMIN,
     *                 XMIN,BESSJO(XMIN),BESSJ1(XMIN)
            ENDIF
          ENDIF
```

```
12    CONTINUE
      END
      FUNCTION FUNC(X)
      FUNC=BESSJO(X)
      RETURN
      END

      PROGRAM D1OR4
C     Driver for routine DBRENT
      PARAMETER(TOL=1.OE-6,EQL=1.E-4)
      EXTERNAL BESSJO,DERIV
      DIMENSION AMIN(20)
      NMIN=0
      WRITE(*,'(/1X,A)') 'Minima of the function BESSJO'
      WRITE(*,'(/1X,T6,A,T19,A,T27,A,T39,A,T53,A/)') 'Min. #','X',
     *         'BESSJO(X)','BESSJ1(X)','DBRENT'
      DO 12 I=1,100
        AX=I
        BX=I+1.0
        CALL MNBRAK(AX,BX,CX,FA,FB,FC,BESSJO)
        DBR=DBRENT(AX,BX,CX,BESSJO,DERIV,TOL,XMIN)
        IF (NMIN.EQ.0) THEN
          AMIN(1)=XMIN
          NMIN=1
          WRITE(*,'(1X,5X,I2,3X,4F12.6)') NMIN,XMIN,
     *              BESSJO(XMIN),DERIV(XMIN),DBR
        ELSE
          IFLAG=0
          DO 11 J=1,NMIN
            IF (ABS(XMIN-AMIN(J)).LE.EQL*XMIN) IFLAG=1
11        CONTINUE
          IF (IFLAG.EQ.0) THEN
            NMIN=NMIN+1
            AMIN(NMIN)=XMIN
            WRITE(*,'(1X,5X,I2,3X,4F12.6)') NMIN,XMIN,
     *                BESSJO(XMIN),DERIV(XMIN),DBR
          ENDIF
        ENDIF
12    CONTINUE
      END
      FUNCTION DERIV(X)
      DERIV=-BESSJ1(X)
      END
```

Numerical Recipes presents several methods for minimization in multiple dimensions. Among these, the downhill simplex method carried out by AMOEBA was the only one that did not treat the problem as a series of one-dimensional minimizations. As input, AMOEBA requires the coordinates of $N+1$ vertices of a starting simplex in N-dimensional space, and the values Y of the function at each of these vertices. Sample program D10R5 tries the method out on the exotic function

$$\mathtt{FAMOEB} = 0.6 - J_0[(x-0.5)^2 + (y-0.6)^2 + (z-0.7)^2]$$

which has a minimum at $(x, y, z) = (0.5, 0.6, 0.7)$. As vertices of the starting simplex, specified by P in a DATA statement, we used $(0,0,0)$, $(1,0,0)$, $(0,1,0)$, and $(0,0,1)$. A vector X(I) is set successively to each vertex to allow the evaluation of function values Y. This data is submitted to AMOEBA along with FTOL=1.0E-6 to specify the tolerance on the function value. The vertices and corresponding function values of the final simplex are printed out, and you can easily check whether the specified tolerance is met.

```
      PROGRAM D10R5
C     Driver for routine AMOEBA
      EXTERNAL FAMOEB
      PARAMETER(NP=3,MP=4,FTOL=1.0E-6)
      DIMENSION P(MP,NP),X(NP),Y(MP)
      DATA P/0.0,1.0,0.0,0.0,0.0,0.0,1.0,0.0,0.0,0.0,0.0,1.0/
      NDIM=NP
      DO 12 I=1,MP
        DO 11 J=1,NP
          X(J)=P(I,J)
11      CONTINUE
        Y(I)=FAMOEB(X)
12    CONTINUE
      CALL AMOEBA(P,Y,MP,NP,NDIM,FTOL,FAMOEB,ITER)
      WRITE(*,'(/1X,A,I3)') 'Iterations: ',ITER
      WRITE(*,'(/1X,A)') 'Vertices of final 3-D simplex and'
      WRITE(*,'(1X,A)') 'function values at the vertices:'
      WRITE(*,'(/3X,A,T11,A,T23,A,T35,A,T45,A/)') 'I',
     *          'X(I)','Y(I)','Z(I)','FUNCTION'
      DO 13 I=1,MP
        WRITE(*,'(1X,I3,4F12.6)') I,(P(I,J),J=1,NP),Y(I)
13    CONTINUE
      WRITE(*,'(/1X,A)') 'True minimum is at (0.5,0.6,0.7)'
      END

      FUNCTION FAMOEB(X)
      DIMENSION X(3)
      FAMOEB=0.6-BESSJ0((X(1)-0.5)**2+(X(2)-0.6)**2+(X(3)-0.7)**2)
      END
```

POWELL carries out one-dimensional minimizations along favorable directions in N-dimensional space. The function minimized must be called FUNC, and in sample program D10R6 a function subroutine is defined for

$$\text{FUNC}(x,y,z) = \frac{1}{2} - J_0[(x-1)^2 + (y-2)^2 + (z-3)^2].$$

The program provides POWELL with a starting point P of $(3/2, 3/2, 5/2)$ and a set of initial directions, here chosen to be the unit directions $(1,0,0)$, $(0,1,0)$, and $(0,0,1)$. POWELL performs its one-dimensional minimizations with LINMIN, which is discussed next.

```
      PROGRAM D10R6
C     Driver for routine POWELL
      PARAMETER(NDIM=3,FTOL=1.0E-6)
      DIMENSION P(NDIM),XI(NDIM,NDIM)
      NP=NDIM
      DATA XI/1.0,0.0,0.0,0.0,1.0,0.0,0.0,0.0,1.0/
      DATA P/1.5,1.5,2.5/
      CALL POWELL(P,XI,NDIM,NP,FTOL,ITER,FRET)
      WRITE(*,'(/1X,A,I3)') 'Iterations:',ITER
      WRITE(*,'(/1X,A/1X,3F12.6)') 'Minimum found at: ',(P(I),I=1,NDIM)
      WRITE(*,'(/1X,A,F12.6)') 'Minimum function value =',FRET
      WRITE(*,'(/1X,A)') 'True minimum of function is at:'
```

```
      WRITE(*,'(1X,3F12.6/)') 1.0,2.0,3.0
      END
      FUNCTION FUNC(X)
      DIMENSION X(3)
      FUNC=0.5-BESSJ0((X(1)-1.0)**2+(X(2)-2.0)**2+(X(3)-3.0)**2)
      END
```

LINMIN, as we have said, finds the minimum of a function FUNC along a direction in N-dimensional space. To use it we specify a point P and a direction vector XI, both in N-space. LINMIN then does the book-keeping required to treat the function as a function of position along this line, and minimizes the function with a conventional one-dimensional minimization routine. Sample program D10R7 feeds LINMIN the function

$$\text{FUNC}(x,y,z) = (x-1)^2 + (y-1)^2 + (z-1)^2$$

which has a minimum at $(x,y,z) = (1,1,1)$. It also chooses point P to be the origin $(0,0,0)$, and tries a series of directions

$$\left(\sqrt{2}\cos\left(\frac{\pi}{2}\frac{\text{I}}{10.0}\right),\quad \sqrt{2}\sin\left(\frac{\pi}{2}\frac{\text{I}}{10.0}\right),\quad 1.0\right) \qquad \text{I} = 1,\ldots,10$$

For each pass, the location of the minimum, and the value of the function at the minimum, are printed. Among the directions searched is the direction $(1,1,1)$. Along this direction, of course, the minimum function value should be zero and should occur at $(1,1,1)$.

```
      PROGRAM D10R7
C     Driver for routine LINMIN
      PARAMETER(NDIM=3,PIO2=1.5707963)
      DIMENSION P(NDIM),XI(NDIM)
      WRITE(*,'(/1X,A)') 'Minimum of a 3-D quadratic centered'
      WRITE(*,'(1X,A)') 'at (1.0,1.0,1.0). Minimum is found'
      WRITE(*,'(1X,A)') 'along a series of radials.'
      WRITE(*,'(/1X,T10,A,T22,A,T34,A,T42,A/)') 'x','y','z','minimum'
      DO 11 I=0,10
        X=PIO2*I/10.0
        SR2=SQRT(2.0)
        XI(1)=SR2*COS(X)
        XI(2)=SR2*SIN(X)
        XI(3)=1.0
        P(1)=0.0
        P(2)=0.0
        P(3)=0.0
        CALL LINMIN(P,XI,NDIM,FRET)
        WRITE(*,'(1X,4F12.6)') (P(J),J=1,3),FRET
11    CONTINUE
      END
      FUNCTION FUNC(X)
      DIMENSION X(3)
      FUNC=0.0
      DO 11 I=1,3
        FUNC=FUNC+(X(I)-1.0)**2
11    CONTINUE
      END
```

F1DIM accompanies LINMIN and is the routine that makes the N-dimensional function FUNC effectively a one-dimensional function along a given line in N-space. There is little to check here, and our perfunctory demonstration of its use, in sample program D10R8, simply plots F1DIM as a one dimensional function, given the function

$$\text{FUNC}(x, y, z) = (x-1)^2 + (y-1)^2 + (z-1)^2.$$

You get to choose the direction; then SCRSHO plots the function along this direction. Try the direction $(1, 1, 1)$ along which you should find a minimum value of FUNC=0 at position $(1, 1, 1)$.

```
      PROGRAM D10R8
C     Driver for routine F1DIM
      PARAMETER(NDIM=3,NMAX=50)
      EXTERNAL F1DIM
      COMMON /F1COM/ NCOM,PCOM(NMAX),XICOM(NMAX)
      DIMENSION P(NDIM),XI(NDIM)
      DATA P/0.0,0.0,0.0/
      NCOM=NDIM
      WRITE(*,'(/1X,A)') 'Enter vector direction along which to'
      WRITE(*,'(1X,A)') 'plot the function. Minimum is in the'
      WRITE(*,'(1X,A)') 'direction 1.0,1.0,1.0 - Enter X,Y,Z:'
      READ(*,*) (XI(I),I=1,3)
      DO 11 J=1,NDIM
        PCOM(J)=P(J)
        XICOM(J)=XI(J)
11    CONTINUE
      CALL SCRSHO(F1DIM)
      END
      FUNCTION FUNC(X)
      DIMENSION X(3)
      FUNC=0.0
      DO 11 I=1,3
        FUNC=FUNC+(X(I)-1.0)**2
11    CONTINUE
      END
```

FRPRMN is another multidimensional minimizer that relies on the one-dimensional minimizations of LINMIN. It works, however, via the Fletcher-Reeves-Polak-Ribiere method and requires that routines be supplied for calculating both the function and its gradient. Sample program D10R9, for example, uses

$$\text{FUNC}(x, y, z) = 1.0 - J_0(x - \tfrac{1}{2}) J_0(y - \tfrac{1}{2}) J_0(z - \tfrac{1}{2})$$

and

$$\frac{\partial\,\text{FUNC}}{\partial x} = J_1(x - \tfrac{1}{2}) J_0(y - \tfrac{1}{2}) J_0(z - \tfrac{1}{2})$$

etc. A number of trial starting vectors are used, and each time, FRPRMN manages to find the minimum at $(1/2, 1/2, 1/2)$.

```
      PROGRAM D10R9
C     Driver for routine FRPRMN
      PARAMETER(NDIM=3,FTOL=1.0E-6,PIO2=1.5707963)
      DIMENSION P(NDIM)
      WRITE(*,'(/1X,A)') 'Program finds the minimum of a function'
      WRITE(*,'(1X,A)') 'with different trial starting vectors.'
```

```
      WRITE(*,'(1X,A)') 'True minimum is (0.5,0.5,0.5)'
      DO 11 K=0,4
        ANGL=PIO2*K/4.0
        P(1)=2.0*COS(ANGL)
        P(2)=2.0*SIN(ANGL)
        P(3)=0.0
        WRITE(*,'(/1X,A,3(F6.4,A))') 'Starting vector: (',
     *          P(1),',',P(2),',',P(3),')'
        CALL FRPRMN(P,NDIM,FTOL,ITER,FRET)
        WRITE(*,'(1X,A,I3)') 'Iterations:',ITER
        WRITE(*,'(1X,A,3(F6.4,A))') 'Solution vector: (',
     *          P(1),',',P(2),',',P(3),')'
        WRITE(*,'(1X,A,E14.6)') 'Func. value at solution',FRET
11    CONTINUE
      END
      FUNCTION FUNC(X)
      DIMENSION X(3)
      FUNC=1.0-BESSJ0(X(1)-0.5)*BESSJ0(X(2)-0.5)*BESSJ0(X(3)-0.5)
      END
      SUBROUTINE DFUNC(X,DF)
      PARAMETER (NMAX=50)
      DIMENSION X(3),DF(NMAX)
      DF(1)=BESSJ1(X(1)-0.5)*BESSJ0(X(2)-0.5)*BESSJ0(X(3)-0.5)
      DF(2)=BESSJ0(X(1)-0.5)*BESSJ1(X(2)-0.5)*BESSJ0(X(3)-0.5)
      DF(3)=BESSJ0(X(1)-0.5)*BESSJ0(X(2)-0.5)*BESSJ1(X(3)-0.5)
      RETURN
      END
```

Completeness requires that we provide a sample program for DF1DIM, which is presented in *Numerical Recipes* as a routine for converting the N-dimensional gradient subroutine to one that provides the first derivative of the function along a specified line in N-dimensional space. It is exactly analogous to F1DIM and the program D1OR10 is the same.

```
      PROGRAM D10R10
C     Driver for routine DF1DIM
      PARAMETER(NDIM=3,NMAX=50)
      EXTERNAL DF1DIM
      COMMON /F1COM/ NCOM,PCOM(NMAX),XICOM(NMAX)
      DIMENSION P(NDIM),XI(NDIM)
      DATA P/0.0,0.0,0.0/
      NCOM=NDIM
      WRITE(*,'(/1X,A)') 'Enter vector direction along which to'
      WRITE(*,'(1X,A)') 'plot the function. Minimum is in the'
      WRITE(*,'(1X,A)') 'direction 1.0,1.0,1.0 - Enter X,Y,Z:'
      READ (*,*) (XI(I),I=1,3)
      DO 11 J=1,NDIM
        PCOM(J)=P(J)
        XICOM(J)=XI(J)
11    CONTINUE
      CALL SCRSHO(DF1DIM)
      END
      SUBROUTINE DFUNC(X,DF)
      DIMENSION X(3),DF(3)
      DO 11 I=1,3
        DF(I)=(X(I)-1.0)**2
11    CONTINUE
```

```
      RETURN
      END
```

DFPMIN implements the Broyden-Fletcher-Goldfarb-Shanno variant of the Davidon-Fletcher-Powell minimization by variable metric methods. It requires somewhat more intermediate storage than the preceding routine and is not considered superior in other ways. However, it is a popular method. Sample program D10R11 works just as did the program for FRPRMN, including the fact that it requires a subroutine for calculation of the derivative.

```
      PROGRAM D10R11
C     Driver for routine DFPMIN
      PARAMETER(NDIM=3,FTOL=1.0E-6,PIO2=1.5707963)
      DIMENSION P(NDIM)
      WRITE(*,'(/1X,A)') 'Program finds the minimum of a function'
      WRITE(*,'(1X,A)') 'with different trial starting vectors.'
      WRITE(*,'(1X,A)') 'True minimum is (0.5,0.5,0.5)'
      DO 11 K=0,4
        ANGL=PIO2*K/4.0
        P(1)=2.0*COS(ANGL)
        P(2)=2.0*SIN(ANGL)
        P(3)=0.0
        WRITE(*,'(/1X,A,3(F6.4,A))') 'Starting vector: (',
     *         P(1),',',P(2),',',P(3),')'
        CALL DFPMIN(P,NDIM,FTOL,ITER,FRET)
        WRITE(*,'(1X,A,I3)') 'Iterations:',ITER
        WRITE(*,'(1X,A,3(F6.4,A))') 'Solution vector: (',
     *         P(1),',',P(2),',',P(3),')'
        WRITE(*,'(1X,A,E14.6)') 'Func. value at solution',FRET
11    CONTINUE
      END
      FUNCTION FUNC(X)
      DIMENSION X(3)
      FUNC=1.0-BESSJ0(X(1)-0.5)*BESSJ0(X(2)-0.5)*BESSJ0(X(3)-0.5)
      END
      SUBROUTINE DFUNC(X,DF)
      PARAMETER (NMAX=50)
      DIMENSION X(3),DF(NMAX)
      DF(1)=BESSJ1(X(1)-0.5)*BESSJ0(X(2)-0.5)*BESSJ0(X(3)-0.5)
      DF(2)=BESSJ0(X(1)-0.5)*BESSJ1(X(2)-0.5)*BESSJ0(X(3)-0.5)
      DF(3)=BESSJ0(X(1)-0.5)*BESSJ0(X(2)-0.5)*BESSJ1(X(3)-0.5)
      RETURN
      END
```

SIMPLX is a subroutine for dealing with problems in linear programming. In these problems the goal is to maximize a linear combination of N variables, subject to the constraint that none be negative, and that as a group they satisfy a number of other constraints. In order to clarify the subject, *Numerical Recipes* presents a sample problem in equations (10.8.6) and (10.8.7), translating the problem into tableau format in (10.8.18), and presenting a solution in equation (10.8.19). Sample program D10R12 carries out the analysis that leads to this solution.

```
      PROGRAM D10R12
C     Driver for routine SIMPLX
C     Incorporates examples discussed in text
      PARAMETER(N=4,M=4,NP=5,MP=6,M1=2,M2=1,M3=1,NM1M2=N+M1+M2)
```

```
      DIMENSION A(MP,NP),IZROV(N),IPOSV(M),ANUM(NP)
      CHARACTER TXT(NM1M2)*2,ALPHA(NP)*2
      DATA TXT/'x1','x2','x3','x4','y1','y2','y3'/
      DATA A/0.0,740.0,0.0,0.5,9.0,0.0,1.0,-1.0,0.0,0.0,-1.0,0.0,
     *        1.0,0.0,-2.0,-1.0,-1.0,0.0,3.0,-2.0,0.0,1.0,-1.0,0.0,
     *        -0.5,0.0,7.0,-2.0,-1.0,0.0/
      CALL SIMPLX(A,M,N,MP,NP,M1,M2,M3,ICASE,IZROV,IPOSV)
      IF (ICASE.EQ.1) THEN
          WRITE(*,*) 'Unbounded objective function'
      ELSE IF (ICASE.EQ.-1) THEN
          WRITE(*,*) 'No solutions satisfy constraints given'
      ELSE
          JJ=1
          DO 11 I=1,N
              IF (IZROV(I).LE.(N+M1+M2)) THEN
                  ALPHA(JJ)=TXT(IZROV(I))
                  JJ=JJ+1
              ENDIF
11        CONTINUE
          JMAX=JJ-1
          WRITE(*,'(/3X,5A10)') '  ',(ALPHA(JJ),JJ=1,JMAX)
          DO 13 I=1,M+1
              IF (I.GT.1) THEN
                  ALPHA(1)=TXT(IPOSV(I-1))
              ELSE
                  ALPHA(1)='  '
              ENDIF
              ANUM(1)=A(I,1)
              JJ=2
              DO 12 J=2,N+1
                  IF (IZROV(J-1).LE.(N+M1+M2)) THEN
                      ANUM(JJ)=A(I,J)
                      JJ=JJ+1
                  ENDIF
12            CONTINUE
              JMAX=JJ-1
              WRITE(*,'(1X,A3,(5F10.2))') ALPHA(1),
     *                (ANUM(JJ),JJ=1,JMAX)
13        CONTINUE
      ENDIF
      END
```

ANNEAL is a subroutine for solving the travelling salesman problem—a problem that is included as a demonstration of the use of simulated annealing. Sample program D10R13 has the function of setting up the initial route for the salesman and printing final results. For each of NCITY=10 cities, it chooses random coordinates X(I),Y(I) using routine RAN3, and puts an entry for each city in the array IPTR(I). The array indicates the order in which the cities will be visited. On the originally specified path, the cities are in the order I=1,...,10 so the sample program initially takes IPTR(I)=I. (It is assumed that the salesman will return to the first city after visiting the last.) A call is then made to ANNEAL, which attempts to find the shortest alternative route, which is recorded in the array. After finding a path that resists further improvement, the driver lists the modified itinerary.

```
      PROGRAM D10R13
      PARAMETER (NCITY=10)
      DIMENSION X(NCITY),Y(NCITY),IORDER(NCITY)
C     Create points of sale
      IDUM=-111
      DO I=1,NCITY
          X(I)=RAN3(IDUM)
          Y(I)=RAN3(IDUM)
          IORDER(I)=I
      ENDDO
      CALL ANNEAL(X,Y,IORDER,NCITY)
      WRITE(*,*) '*** System Frozen ***'
      WRITE(*,*) 'Final path:'
      WRITE(*,'(1X,T3,A,T13,A,T23,A)') 'city','x','y'
      DO I=1,NCITY
          II=IORDER(I)
          WRITE(*,'(1X,I4,2F10.4)') II,X(II),Y(II)
      ENDDO
      END
```

Chapter 11: Eigensystems

In Chapter 11 of Numerical Recipes, we deal with the problem of finding eigenvectors and eigenvalues of matrices, first dealing with symmetric matrices, and then with more general cases. For real symmetric matrices of small-to-moderate size, the routine `JACOBI` *is recommended as a simple and foolproof scheme of finding eigenvalues and eigenvectors. Routine* `EIGSRT` *may be used to reorder the output of* `JACOBI` *into ascending order of eigenvalue. A more efficient (but operationally more complicated) procedure is to reduce the symmetric matrix to tridiagonal form before doing the eigenvalue analysis.* `TRED2` *uses the Householder scheme to perform this reduction and is used in conjunction with* `TQLI`*.* `TQLI` *determines the eigenvalues and eigenvectors of a real, symmetric, tridiagonal matrix.*

For nonsymmetric matrices, we offer only routines for finding eigenvalues, and not eigenvectors. To ameliorate problems with roundoff error, `BALANC` *makes the corresponding rows and columns of the matrix have comparable norms while leaving eigenvalues unchanged. Then the matrix is reduced to Hessenberg form by Gaussian elimination using* `ELMHES`*. Finally* `HQR` *applies the QR algorithm to find the eigenvalues of the Hessenberg matrix.*

⋆ ⋆ ⋆ ⋆

`JACOBI` is a reliable scheme for finding both the eigenvalues and eigenvectors of a symmetric matrix. It is not the most efficient scheme available, but it is simple and trustworthy, and it is recommended for problems of small-to-moderate order. Sample program `D11R1` defines three matrices `A,B,C` for use by `JACOBI`. They are of order 3, 5, and 10 respectively. To check the index handling in `JACOBI`, they are, each in turn, loaded into a 10×10 matrix `E`, and then sent to `JACOBI` with `NP=10` (the physical size of the matrix) and `N=3, 5 or 10` (the logical array dimension). For each matrix, the eigenvalues and eigenvectors are reported. Then, an eigenvector test takes place in which the original matrix is applied to the purported eigenvector, and the ratio of the result to the vector itself is found. The ratio should, of course, be the eigenvalue.

```
      PROGRAM D11R1
C     Driver for routine JACOBI
      PARAMETER(NP=10,NMAT=3)
      DIMENSION D(NP),V(NP,NP),R(NP)
      DIMENSION A(3,3),B(5,5),C(10,10),E(NP,NP),NUM(3)
      DATA NUM/3,5,10/
      DATA A/1.0,2.0,3.0,2.0,2.0,3.0,3.0,3.0,3.0/
      DATA B/-2.0,-1.0,0.0,1.0,2.0,-1.0,-1.0,0.0,1.0,2.0,
```

```
     *         0.0,0.0,0.0,1.0,2.0,1.0,1.0,1.0,1.0,2.0,
     *         2.0,2.0,2.0,2.0,2.0/
      DATA C/5.0,4.0,3.0,2.0,1.0,0.0,-1.0,-2.0,-3.0,-4.0,
     *         4.0,5.0,4.0,3.0,2.0,1.0,0.0,-1.0,-2.0,-3.0,
     *         3.0,4.0,5.0,4.0,3.0,2.0,1.0,0.0,-1.0,-2.0,
     *         2.0,3.0,4.0,5.0,4.0,3.0,2.0,1.0,0.0,-1.0,
     *         1.0,2.0,3.0,4.0,5.0,4.0,3.0,2.0,1.0,0.0,
     *         0.0,1.0,2.0,3.0,4.0,5.0,4.0,3.0,2.0,1.0,
     *         -1.0,0.0,1.0,2.0,3.0,4.0,5.0,4.0,3.0,2.0,
     *         -2.0,-1.0,0.0,1.0,2.0,3.0,4.0,5.0,4.0,3.0,
     *         -3.0,-2.0,-1.0,0.0,1.0,2.0,3.0,4.0,5.0,4.0,
     *         -4.0,-3.0,-2.0,-1.0,0.0,1.0,2.0,3.0,4.0,5.0/
      DO 24 I=1,NMAT
          IF (I.EQ.1) THEN
              DO 12 II=1,3
                  DO 11 JJ=1,3
                      E(II,JJ)=A(II,JJ)
11                CONTINUE
12            CONTINUE
              CALL JACOBI(E,3,NP,D,V,NROT)
          ELSE IF (I.EQ.2) THEN
              DO 14 II=1,5
                  DO 13 JJ=1,5
                      E(II,JJ)=B(II,JJ)
13                CONTINUE
14            CONTINUE
              CALL JACOBI(E,5,NP,D,V,NROT)
          ELSE IF (I.EQ.3) THEN
              DO 16 II=1,10
                  DO 15 JJ=1,10
                      E(II,JJ)=C(II,JJ)
15                CONTINUE
16            CONTINUE
              CALL JACOBI(E,10,NP,D,V,NROT)
          ENDIF
          WRITE(*,'(/1X,A,I2)') 'Matrix Number',I
          WRITE(*,'(1X,A,I3)') 'Number of JACOBI rotations:',NROT
          WRITE(*,'(/1X,A)') 'Eigenvalues:'
          DO 17 J=1,NUM(I)
              WRITE(*,'(1X,5F12.6)') D(J)
17        CONTINUE
          WRITE(*,'(/1X,A)') 'Eigenvectors:'
          DO 18 J=1,NUM(I)
              WRITE(*,'(1X,T5,A,I3)') 'Number',J
              WRITE(*,'(1X,5F12.6)') (V(K,J),K=1,NUM(I))
18        CONTINUE
C     Eigenvector test
          WRITE(*,'(/1X,A)') 'Eigenvector Test'
          DO 23 J=1,NUM(I)
              DO 21 L=1,NUM(I)
                  R(L)=0.0
                  DO 19 K=1,NUM(I)
                      IF (K.GT.L) THEN
                          KK=L
                          LL=K
                      ELSE
                          KK=K
```

```
                        LL=L
                    ENDIF
                    IF (I.EQ.1) THEN
                        R(L)=R(L)+A(LL,KK)*V(K,J)
                    ELSE IF (I.EQ.2) THEN
                        R(L)=R(L)+B(LL,KK)*V(K,J)
                    ELSE IF (I.EQ.3) THEN
                        R(L)=R(L)+C(LL,KK)*V(K,J)
                    ENDIF
19              CONTINUE
21          CONTINUE
            WRITE(*,'(/1X,A,I3)') 'Vector Number',J
            WRITE(*,'(/1X,T7,A,T18,A,T31,A)')
     *              'Vector','Mtrx*Vec.','Ratio'
            DO 22 L=1,NUM(I)
                RATIO=R(L)/V(L,J)
                WRITE(*,'(1X,3F12.6)') V(L,J),R(L),RATIO
22          CONTINUE
23      CONTINUE
        WRITE(*,*) 'press RETURN to continue...'
        READ(*,*)
24    CONTINUE
      END
```

EIGSRT reorders the output of JACOBI so that the eigenvectors are in the order of increasing eigenvalue. Sample program D11R2 uses matrix C from the previous program to illustrate. This 10×10 matrix is passed to JACOBI and the ten eigenvectors are found. They are printed, along with their eigenvalues, in the order that JACOBI returns them. Then the matrices D and V from JACOBI, which contain the eigenvalues and eigenvectors, are passed to EIGSRT, and ought to return in ascending order of eigenvalue. The result is printed for inspection.

```
      PROGRAM D11R2
C     Driver for routine EIGSRT
      PARAMETER(NP=10)
      DIMENSION D(NP),V(NP,NP),C(NP,NP)
      DATA C /5.0,4.0,3.0,2.0,1.0,0.0,-1.0,-2.0,-3.0,-4.0,
     *        4.0,5.0,4.0,3.0,2.0,1.0,0.0,-1.0,-2.0,-3.0,
     *        3.0,4.0,5.0,4.0,3.0,2.0,1.0,0.0,-1.0,-2.0,
     *        2.0,3.0,4.0,5.0,4.0,3.0,2.0,1.0,0.0,-1.0,
     *        1.0,2.0,3.0,4.0,5.0,4.0,3.0,2.0,1.0,0.0,
     *        0.0,1.0,2.0,3.0,4.0,5.0,4.0,3.0,2.0,1.0,
     *        -1.0,0.0,1.0,2.0,3.0,4.0,5.0,4.0,3.0,2.0,
     *        -2.0,-1.0,0.0,1.0,2.0,3.0,4.0,5.0,4.0,3.0,
     *        -3.0,-2.0,-1.0,0.0,1.0,2.0,3.0,4.0,5.0,4.0,
     *        -4.0,-3.0,-2.0,-1.0,0.0,1.0,2.0,3.0,4.0,5.0/
      CALL JACOBI(C,NP,NP,D,V,NROT)
      WRITE(*,*) 'Unsorted Eigenvectors:'
      DO 11 I=1,NP
        WRITE(*,'(/1X,A,I3,A,F12.6)') 'Eigenvalue',I,' =',D(I)
        WRITE(*,*) 'Eigenvector:'
        WRITE(*,'(10X,5F12.6)') (V(J,I),J=1,NP)
11    CONTINUE
      WRITE(*,'(//,A,//)') '****** sorting ******'
      CALL EIGSRT(D,V,NP,NP)
      WRITE(*,*) 'Sorted Eigenvectors:'
```

```
      DO 12 I=1,NP
          WRITE(*,'(/1X,A,I3,A,F12.6)') 'Eigenvalue',I,' =',D(I)
          WRITE(*,*) 'Eigenvector:'
          WRITE(*,'(10X,5F12.6)') (V(J,I),J=1,NP)
12    CONTINUE
      END
```

TRED2 reduces a real symmetric matrix to tridiagonal form. Sample program TRED2 again uses matrix C from the earlier programs, and copies it into matrix A. Matrix A is sent to TRED2, while C is saved for a check of the transformation matrix that TRED2 returns in A. The program prints the diagonal and off-diagonal elements of the reduced matrix. It then forms the matrix F defined by $F = A^T C A$ to prove that F is tridiagonal and that the listed diagonal and off-diagonal elements are correct.

```
      PROGRAM D11R3
C     Driver for routine TRED2
      PARAMETER(NP=10)
      DIMENSION A(NP,NP),C(NP,NP),D(NP),E(NP),F(NP,NP)
      DATA C/5.0,4.0,3.0,2.0,1.0,0.0,-1.0,-2.0,-3.0,-4.0,
     *        4.0,5.0,4.0,3.0,2.0,1.0,0.0,-1.0,-2.0,-3.0,
     *        3.0,4.0,5.0,4.0,3.0,2.0,1.0,0.0,-1.0,-2.0,
     *        2.0,3.0,4.0,5.0,4.0,3.0,2.0,1.0,0.0,-1.0,
     *        1.0,2.0,3.0,4.0,5.0,4.0,3.0,2.0,1.0,0.0,
     *        0.0,1.0,2.0,3.0,4.0,5.0,4.0,3.0,2.0,1.0,
     *        -1.0,0.0,1.0,2.0,3.0,4.0,5.0,4.0,3.0,2.0,
     *        -2.0,-1.0,0.0,1.0,2.0,3.0,4.0,5.0,4.0,3.0,
     *        -3.0,-2.0,-1.0,0.0,1.0,2.0,3.0,4.0,5.0,4.0,
     *        -4.0,-3.0,-2.0,-1.0,0.0,1.0,2.0,3.0,4.0,5.0/
      DO 12 I=1,NP
          DO 11 J=1,NP
              A(I,J)=C(I,J)
11        CONTINUE
12    CONTINUE
      CALL TRED2(A,NP,NP,D,E)
      WRITE(*,'(/1X,A)') 'Diagonal elements'
      WRITE(*,'(1X,5F12.6)') (D(I),I=1,NP)
      WRITE(*,'(/1X,A)') 'Off-diagonal elements'
      WRITE(*,'(1X,5F12.6)') (E(I),I=2,NP)
C     Check transformation matrix
      DO 16 J=1,NP
          DO 15 K=1,NP
          F(J,K)=0.0
              DO 14 L=1,NP
                  DO 13 M=1,NP
                      F(J,K)=F(J,K)
     *                    +A(L,J)*C(L,M)*A(M,K)
13                CONTINUE
14            CONTINUE
15        CONTINUE
16    CONTINUE
C     How does it look?
      WRITE(*,'(/1X,A)') 'Tridiagonal matrix'
      DO 17 I=1,NP
          WRITE(*,'(1X,10F7.2)') (F(I,J),J=1,NP)
17    CONTINUE
```

```
      END
```

TQLI finds the eigenvectors and eigenvalues for a real, symmetric, tridiagonal matrix. Sample program D11R4 operates with matrix C again, and uses TRED2 to reduce it to tridiagonal form as before. More specifically, C is copied into matrix A, which is sent to TRED2. From TRED2 come two vectors D,E which are the diagonal and subdiagonal elements of the tridiagonal matrix. D and E are made arguments of TQLI, as is A, the returned transformation matrix from TRED2. On output from TQLI, D is replaced with eigenvalues, and A with corresponding eigenvectors. These are checked as in the program for JACOBI. That is, the original matrix C is applied to each eigenvector, and the result is divided (element by element) by the eigenvector. Look for a result equal to the eigenvalue. (Note: in some cases, the vector element is zero or nearly so. These cases are flagged with the words "div. by zero".)

```
      PROGRAM D11R4
C     Driver for routine TQLI
      PARAMETER(NP=10,TINY=1.0E-6)
      DIMENSION A(NP,NP),C(NP,NP),D(NP),E(NP),F(NP)
      DATA C/5.0,4.0,3.0,2.0,1.0,0.0,-1.0,-2.0,-3.0,-4.0,
     *       4.0,5.0,4.0,3.0,2.0,1.0,0.0,-1.0,-2.0,-3.0,
     *       3.0,4.0,5.0,4.0,3.0,2.0,1.0,0.0,-1.0,-2.0,
     *       2.0,3.0,4.0,5.0,4.0,3.0,2.0,1.0,0.0,-1.0,
     *       1.0,2.0,3.0,4.0,5.0,4.0,3.0,2.0,1.0,0.0,
     *       0.0,1.0,2.0,3.0,4.0,5.0,4.0,3.0,2.0,1.0,
     *       -1.0,0.0,1.0,2.0,3.0,4.0,5.0,4.0,3.0,2.0,
     *       -2.0,-1.0,0.0,1.0,2.0,3.0,4.0,5.0,4.0,3.0,
     *       -3.0,-2.0,-1.0,0.0,1.0,2.0,3.0,4.0,5.0,4.0,
     *       -4.0,-3.0,-2.0,-1.0,0.0,1.0,2.0,3.0,4.0,5.0/
      DO 12 I=1,NP
        DO 11 J=1,NP
          A(I,J)=C(I,J)
11      CONTINUE
12    CONTINUE
      CALL TRED2(A,NP,NP,D,E)
      CALL TQLI(D,E,NP,NP,A)
      WRITE(*,'(/1X,A)') 'Eigenvectors for a real symmetric matrix'
      DO 16 I=1,NP
        DO 14 J=1,NP
          F(J)=0.0
          DO 13 K=1,NP
            F(J)=F(J)+C(J,K)*A(K,I)
13        CONTINUE
14      CONTINUE
        WRITE(*,'(/1X,A,I3,A,F10.6)') 'Eigenvalue',I,' =',D(I)
        WRITE(*,'(/1X,T7,A,T17,A,T31,A)') 'Vector','Mtrx*Vect.','Ratio'
        DO 15 J=1,NP
          IF (ABS(A(J,I)).LT.TINY) THEN
            WRITE(*,'(1X,2F12.6,A12)') A(J,I),F(J),'div. by 0'
          ELSE
            WRITE(*,'(1X,2F12.6,E14.6)') A(J,I),F(J),
     *              F(J)/A(J,I)
          ENDIF
15      CONTINUE
        WRITE(*,'(/1X,A)') 'press ENTER to continue...'
        READ(*,*)
16    CONTINUE
```

```
      END
```

BALANC reduces error in eigenvalue problems involving non-symmetric matrices. It does this by adjusting corresponding rows and columns to have comparable norms, without changing eigenvalues. Sample program D11R5 prepares the following array A for BALANC

$$\begin{pmatrix} 1 & 100 & 1 & 100 & 1 \\ 1 & 1 & 1 & 1 & 1 \\ 1 & 100 & 1 & 100 & 1 \\ 1 & 1 & 1 & 1 & 1 \\ 1 & 100 & 1 & 100 & 1 \end{pmatrix}$$

The norms of the five rows and five columns are printed out. It is clear from the array that three of the rows and two of the columns have much larger norms than the others. After balancing with BALANC, the norms are recalculated, and this time the row, column pairs should be much more nearly equal.

```
      PROGRAM D11R5
C     Driver for routine BALANC
      PARAMETER(NP=5)
      DIMENSION A(NP,NP),R(NP),C(NP)
      DATA A/1.0,1.0,1.0,1.0,1.0,100.0,1.0,100.0,1.0,100.0,
     *       1.0,1.0,1.0,1.0,1.0,100.0,1.0,100.0,1.0,100.0,
     *       1.0,1.0,1.0,1.0,1.0/
C     Print norms
      DO 12 I=1,NP
        R(I)=0.0
        C(I)=0.0
        DO 11 J=1,NP
          R(I)=R(I)+ABS(A(I,J))
          C(I)=C(I)+ABS(A(J,I))
11      CONTINUE
12    CONTINUE
      WRITE(*,*) 'Rows:'
      WRITE(*,*) (R(I),I=1,NP)
      WRITE(*,*) 'Columns:'
      WRITE(*,*) (C(I),I=1,NP)
      WRITE(*,'(/1X,A/)') '***** Balancing Matrix *****'
      CALL BALANC(A,NP,NP)
C     Print norms
      DO 14 I=1,NP
        R(I)=0.0
        C(I)=0.0
        DO 13 J=1,NP
          R(I)=R(I)+ABS(A(I,J))
          C(I)=C(I)+ABS(A(J,I))
13      CONTINUE
14    CONTINUE
      WRITE(*,*) 'Rows:'
      WRITE(*,*) (R(I),I=1,NP)
      WRITE(*,*) 'Columns:'
      WRITE(*,*) (C(I),I=1,NP)
      END
```

ELMHES reduces a general matrix to Hessenberg form using Gaussian elimination.

It is particularly valuable for real, non-symmetric matrices. Sample program D11R6 employs BALANC and ELMHES to get a non-symmetric and grossly unbalanced matrix into Hessenberg form. The matrix A is

$$\begin{pmatrix} 1 & 2 & 300 & 4 & 5 \\ 2 & 3 & 400 & 5 & 6 \\ 3 & 4 & 5 & 6 & 7 \\ 4 & 5 & 600 & 7 & 8 \\ 5 & 6 & 700 & 8 & 9 \end{pmatrix}$$

After printing the original matrix, the program feeds it to BALANC and prints the balanced version. This is submitted to ELMHES and the result is printed. Notice that the elements of A with I>J+1 are all set to zero by the program, because ELMHES returns random values in this part of the matrix. Therefore, you should not attach any importance to the fact that the printed output of the program has Hessenberg form. More important are the contents of the non-zero entries. We include here the expected results for comparison.

Balanced Matrix:

```
       1.00        2.00       37.50        4.00        5.00
       2.00        3.00       50.00        5.00        6.00
      24.00       32.00        5.00       48.00       56.00
       4.00        5.00       75.00        7.00        8.00
       5.00        6.00       87.50        8.00        9.00
```

Reduced to Hessenberg Form:

```
  .1000E+01   .3938E+02   .9618E+01   .3333E+01   .4000E+01
  .2400E+02   .2733E+02   .1161E+03   .4800E+02   .4800E+02
  .0000E+00   .8551E+02  -.4780E+01  -.1333E+01  -.2000E+01
  .0000E+00   .0000E+00   .5188E+01   .1447E+01   .2171E+01
  .0000E+00   .0000E+00   .0000E+00  -.9155E-07   .7874E-07
```

```
      PROGRAM D11R6
C     Driver for ELMHES
      PARAMETER(NP=5)
      DIMENSION A(NP,NP),R(NP),C(NP)
      DATA A/1.0,2.0,3.0,4.0,5.0,2.0,3.0,4.0,5.0,6.0,
     *       300.0,400.0,5.0,600.0,700.0,4.0,5.0,6.0,7.0,8.0,
     *       5.0,6.0,7.0,8.0,9.0/
      WRITE(*,'(/1X,A/)') '***** Original Matrix *****'
      DO 11 I=1,NP
        WRITE(*,'(1X,5F12.2)') (A(I,J),J=1,NP)
11    CONTINUE
      WRITE(*,'(/1X,A/)') '***** Balance Matrix *****'
      CALL BALANC(A,NP,NP)
      DO 12 I=1,NP
        WRITE(*,'(1X,5F12.2)') (A(I,J),J=1,NP)
12    CONTINUE
      WRITE(*,'(/1X,A/)') '***** Reduce to Hessenberg Form *****'
      CALL ELMHES(A,NP,NP)
      DO 14 J=1,NP-2
        DO 13 I=J+2,NP
          A(I,J)=0.0
13      CONTINUE
14    CONTINUE
```

```
      DO 15 I=1,NP
          WRITE(*,'(1X,5E12.4)') (A(I,J),J=1,NP)
15    CONTINUE
      END
```

HQR, finally, is a routine for finding the eigenvalues of a Hessenberg matrix using the QR algorithm. The 5×5 matrix A specified in the DATA statement is treated just as you would expect to treat any general real non-symmetric matrix. It is fed to BALANC for balancing, to ELMHES for reduction to Hessenberg form, and to HQR for eigenvalue determination. The eigenvalues may be complex-valued, and both real and imaginary parts are given. The original matrix has enough strategically placed zeros in it that you should have no trouble finding the eigenvalues by hand. Alternatively, you may check them against the list below:

Matrix:

```
    1.00        2.00         .00         .00         .00
   -2.00        3.00         .00         .00         .00
    3.00        4.00       50.00         .00         .00
   -4.00        5.00      -60.00        7.00         .00
   -5.00        6.00      -70.00        8.00       -9.00
```

Eigenvalues:

```
    #       Real           Imag.
    1     .500000E+02     .000000E+00
    2     .200000E+01    -.173205E+01
    3     .200000E+01     .173205E+01
    4     .700000E+01     .000000E+00
    5    -.900000E+01     .000000E+00
```

```
      PROGRAM D11R7
C     Driver for routine HQR
      PARAMETER(NP=5)
      DIMENSION A(NP,NP),WR(NP),WI(NP)
      DATA A/1.0,-2.0,3.0,-4.0,-5.0,2.0,3.0,4.0,5.0,6.0,
     *        0.0,0.0,50.0,-60.0,-70.0,0.0,0.0,0.0,7.0,8.0,
     *        0.0,0.0,0.0,0.0,-9.0/
      WRITE(*,'(/1X,A)') 'Matrix:'
      DO 11 I=1,NP
          WRITE(*,'(1X,5F12.2)') (A(I,J),J=1,NP)
11    CONTINUE
      CALL BALANC(A,NP,NP)
      CALL ELMHES(A,NP,NP)
      CALL HQR(A,NP,NP,WR,WI)
      WRITE(*,'(/1X,A)') 'Eigenvalues:'
      WRITE(*,'(/1X,T9,A,T24,A/)') 'Real','Imag.'
      DO 12 I=1,NP
          WRITE(*,'(1X,2E15.6)') WR(I),WI(I)
12    CONTINUE
      END
```

Chapter 12: Fourier Methods

Chapter 12 of Numerical Recipes covers Fourier transform spectral methods, particularly the transform of discretely sampled data. Central to the chapter is the fast Fourier transform (FFT). Routine FOUR1 *performs the FFT on a complex data array.* TWOFFT *does the same transform on two real-valued data arrays (at the same time) and returns two complex-valued transforms. Finally,* REALFT *finds the Fourier transform of a single real-valued array. Two related transforms are the sine transform and the cosine transform, given by* SINFT *and* COSFT.

Two common uses of the Fourier transform are the convolution of data with a response function, and the computation of the correlation of two data sets. These operations are carried out by CONVLV *and* CORREL *respectively. Other applications of Fourier methods include data filtering, power spectrum estimation (*SPCTRM*, or* EVLMEM *with* MEMCOF*), and linear prediction (*PREDIC *with* FIXRTS*). All of these applications assume data in one dimension. For FFTs in two or more dimensions the routine* FOURN *is supplied.*

⋆ ⋆ ⋆ ⋆

Routine FOUR1 performs the fast Fourier transform on a complex-valued array of data points. Example program D12R1 has five tests for this transform. First, it checks the following four symmetries (where $h(t)$ is the data and $H(n)$ is the transform):

1. If $h(t)$ is real-valued and even, then $H(n) = H(N - n)$ and H is real.
2. If $h(t)$ is imaginary-valued and even, then $H(n) = H(N-n)$ and H is imaginary.
3. If $h(t)$ is real-valued and odd, then $H(n) = -H(N - n)$ and H is imaginary.
4. If $h(t)$ is imaginary-valued and odd, then $H(n) = -H(N - n)$ and H is real.

The fifth test is that if a data array is Fourier transformed twice in succession, the resulting array should be identical to the original.

```
      PROGRAM D12R1
C     Driver for routine FOUR1
      PARAMETER (NN=32,NN2=2*NN)
      DIMENSION DATA(NN2),DCMP(NN2)
      WRITE(*,*) 'h(t)=real-valued even-function'
      WRITE(*,*) 'H(n)=H(N-n) and real?'
      DO 11 I=1,2*NN-1,2
        DATA(I)=1.0/(((I-NN-1.0)/NN)**2+1.0)
        DATA(I+1)=0.0
11    CONTINUE
```

```
      ISIGN=1
      CALL FOUR1(DATA,NN,ISIGN)
      CALL PRNTFT(DATA,NN2)
      WRITE(*,*) 'h(t)=imaginary-valued even-function'
      WRITE(*,*) 'H(n)=H(N-n) and imaginary?'
      DO 12 I=1,2*NN-1,2
          DATA(I+1)=1.0/(((I-NN-1.0)/NN)**2+1.0)
          DATA(I)=0.0
12    CONTINUE
      ISIGN=1
      CALL FOUR1(DATA,NN,ISIGN)
      CALL PRNTFT(DATA,NN2)
      WRITE(*,*) 'h(t)=real-valued odd-function'
      WRITE(*,*) 'H(n)=-H(N-n) and imaginary?'
      DO 13 I=1,2*NN-1,2
          DATA(I)=(I-NN-1.0)/NN/(((I-NN-1.0)/NN)**2+1.0)
          DATA(I+1)=0.0
13    CONTINUE
      DATA(1)=0.0
      ISIGN=1
      CALL FOUR1(DATA,NN,ISIGN)
      CALL PRNTFT(DATA,NN2)
      WRITE(*,*) 'h(t)=imaginary-valued odd-function'
      WRITE(*,*) 'H(n)=-H(N-n) and real?'
      DO 14 I=1,2*NN-1,2
          DATA(I+1)=(I-NN-1.0)/NN/(((I-NN-1.0)/NN)**2+1.0)
          DATA(I)=0.0
14    CONTINUE
      DATA(2)=0.0
      ISIGN=1
      CALL FOUR1(DATA,NN,ISIGN)
      CALL PRNTFT(DATA,NN2)
C     Transform, inverse-transform test
      DO 15 I=1,2*NN-1,2
          DATA(I)=1.0/((0.5*(I-NN-1)/NN)**2+1.0)
          DCMP(I)=DATA(I)
          DATA(I+1)=(0.25*(I-NN-1)/NN)*
     *              EXP(-(0.5*(I-NN-1.0)/NN)**2)
          DCMP(I+1)=DATA(I+1)
15    CONTINUE
      ISIGN=1
      CALL FOUR1(DATA,NN,ISIGN)
      ISIGN=-1
      CALL FOUR1(DATA,NN,ISIGN)
      WRITE(*,'(/1X,T10,A,T44,A)') 'Double Fourier Transform:',
     *        'Original Data:'
      WRITE(*,'(/1X,T5,A,T11,A,T24,A,T41,A,T53,A/)')
     *        'k','Real h(k)','Imag h(k)','Real h(k)','Imag h(k)'
      DO 16 I=1,NN,2
          J=(I+1)/2
          WRITE(*,'(1X,I4,2X,2F12.6,5X,2F12.6)') J,DCMP(I),
     *              DCMP(I+1),DATA(I)/NN,DATA(I+1)/NN
16    CONTINUE
      END
      SUBROUTINE PRNTFT(DATA,NN2)
      DIMENSION DATA(NN2)
      WRITE(*,'(/1X,T5,A,T11,A,T23,A,T39,A,T52,A)')
```

```
     *         'n','Real H(n)','Imag H(n)','Real H(N-n)','Imag H(N-n)'
        WRITE(*,'(1X,I4,2X,2F12.6,5X,2F12.6)') 0,DATA(1),DATA(2),
     *         DATA(1),DATA(2)
        DO 11 N=3,(NN2/2)+1,2
            M=(N-1)/2
            MM=NN2+2-N
            WRITE(*,'(1X,I4,2X,2F12.6,5X,2F12.6)') M,DATA(N),
     *             DATA(N+1),DATA(MM),DATA(MM+1)
11      CONTINUE
        WRITE(*,'(/1X,A)') ' press RETURN to continue ...'
        READ(*,*)
        RETURN
        END
```

TWOFFT is a routine that performs an efficient FFT of two real arrays at once by packing them into a complex array and transforming with FOUR1. Sample program D12R2 generates two periodic data sets, out of phase with one another, and performs a transform and an inverse transform on each. It will be difficult to judge whether the transform itself gives the right answer, but if the inverse transform gets you back to the easily recognized original, you may be fairly confident that the routine works.

```
        PROGRAM D12R2
C       Driver for routine TWOFFT
        PARAMETER(N=32,N2=2*N,PER=8.0,PI=3.14159)
        DIMENSION DATA1(N),DATA2(N),FFT1(N2),FFT2(N2)
        DO 11 I=1,N
            X=2.0*PI*I/PER
            DATA1(I)=NINT(COS(X))
            DATA2(I)=NINT(SIN(X))
11      CONTINUE
        CALL TWOFFT(DATA1,DATA2,FFT1,FFT2,N)
        WRITE(*,*) 'Fourier transform of first function:'
        CALL PRNTFT(FFT1,N2)
        WRITE(*,*) 'Fourier transform of second function:'
        CALL PRNTFT(FFT2,N2)
C       Invert transform
        ISIGN=-1
        CALL FOUR1(FFT1,N,ISIGN)
        WRITE(*,*) 'Inverted transform = first function:'
        CALL PRNTFT(FFT1,N2)
        CALL FOUR1(FFT2,N,ISIGN)
        WRITE(*,*) 'Inverted transform = second function:'
        CALL PRNTFT(FFT2,N2)
        END
        SUBROUTINE PRNTFT(DATA,N2)
        DIMENSION DATA(N2)
        WRITE(*,'(1X,T7,A,T13,A,T24,A,T35,A,T47,A)')
     *         'n','Real(n)','Imag.(n)','Real(N-n)','Imag.(N-n)'
        WRITE(*,'(1X,I6,4F12.6)') 0,DATA(1),DATA(2),DATA(1),DATA(2)
        DO 11 I=3,(N2/2)+1,2
            M=(I-1)/2
            NN2=N2+2-I
            WRITE(*,'(1X,I6,4F12.6)') M,DATA(I),DATA(I+1),
     *             DATA(NN2),DATA(NN2+1)
11      CONTINUE
```

```
      WRITE(*,'(/1X,A)') ' press RETURN to continue ...'
      READ(*,*)
      RETURN
      END
```

REALFT performs the Fourier transform of a single real-valued data array. Sample routine D12R3 takes this function to be sinusoidal, and allows you to choose the period. After transforming, it simply plots the magnitude of each element of the transform. If the period you choose is a power of two, the transform will be nonzero in a single bin; otherwise there will be leakage to adjacent channels. D12R3 follows every transform by an inverse transform to make sure the original function is recovered.

```
      PROGRAM D12R3
C     Driver for routine REALFT
      PARAMETER(EPS=1.0E-3,NP=32,NPP2=NP+2,WIDTH=50.0,PI=3.14159)
      DIMENSION DATA(NPP2),SIZE(NP)
      N=NP/2
1     WRITE(*,'(1X,A,I2,A)') 'Period of sinusoid in channels (2-',
     *        NP,', OR 0 TO STOP)'
      READ(*,*) PER
      IF(PER.LE.0.)STOP
      DO 11 I=1,NP
        DATA(I)=COS(2.0*PI*(I-1)/PER)
11    CONTINUE
      CALL REALFT(DATA,N,+1)
      BIG=-1.0E10
      DO 12 I=1,N
        SIZE(I)=SQRT(DATA(2*I-1)**2+DATA(2*I)**2)
        IF (I.EQ.1) SIZE(I)=DATA(I)
        IF(SIZE(I).GT.BIG) BIG=SIZE(I)
12    CONTINUE
      SCAL=WIDTH/BIG
      DO 13 I=1,N
        NLIM=SCAL*SIZE(I)+EPS
        WRITE(*,'(1X,I4,1X,60A1)') I,('*',J=1,NLIM+1)
13    CONTINUE
      WRITE(*,*) 'press continue ...'
      READ(*,*)
      CALL REALFT(DATA,N,-1)
      BIG=-1.0E10
      SMALL=1.0E10
      DO 14 I=1,NP
        IF(DATA(I).LT.SMALL) SMALL=DATA(I)
        IF(DATA(I).GT.BIG) BIG=DATA(I)
14    CONTINUE
      SCAL=WIDTH/(BIG-SMALL)
      DO 15 I=1,NP
        NLIM=SCAL*(DATA(I)-SMALL)+EPS
        WRITE(*,'(1X,I4,1X,60A1)') I,
     *        ('*',J=1,NLIM+1)
15    CONTINUE
      GOTO 1
      END
```

SINFT performs a sine-transform of a real-valued array. The necessity for such a transform arises in solution methods for partial differential equations with certain kinds of boundary conditions (see Chapter 17). The sample program D12R4 works exactly as the previous program. Notice that in this program no distinction needs to be made between the transform and its inverse. They are identical.

```
      PROGRAM D12R4
C     Driver for routine SINFT
      PARAMETER(EPS=1.0E-3,NP=16,WIDTH=30.0,PI=3.14159)
      DIMENSION DATA(NP),SIZE(NP)
1     WRITE(*,'(1X,A,I2,A)') 'Period of sinusoid in channels (2-',NP,')'
      READ(*,*) PER
      IF (PER.LE.0.) STOP
      DO 11 I=1,NP
        DATA(I)=SIN(2.0*PI*(I-1)/PER)
11    CONTINUE
      CALL SINFT(DATA,NP)
      BIG=-1.0E10
      SMALL=1.0E10
      DO 12 I=1,NP
        IF (DATA(I).LT.SMALL) SMALL=DATA(I)
        IF (DATA(I).GT.BIG) BIG=DATA(I)
12    CONTINUE
      SCAL=WIDTH/(BIG-SMALL)
      DO 13 I=1,NP
        NLIM=SCAL*(DATA(I)-SMALL)+EPS
        WRITE(*,'(1X,I4,1X,60A1)') I,('*',J=1,NLIM+1)
13    CONTINUE
      WRITE(*,*) 'press continue ...'
      READ(*,*)
      CALL SINFT(DATA,NP)
      BIG=-1.0E10
      SMALL=1.0E10
      DO 14 I=1,NP
        IF(DATA(I).LT.SMALL) SMALL=DATA(I)
        IF(DATA(I).GT.BIG) BIG=DATA(I)
14    CONTINUE
      SCAL=WIDTH/(BIG-SMALL)
      DO 15 I=1,NP
        NLIM=SCAL*(DATA(I)-SMALL)+EPS
        WRITE(*,'(1X,I4,1X,60A1)') I,('*',J=1,NLIM+1)
15    CONTINUE
      GOTO 1
      END
```

COSFT is a companion subroutine to SINFT that does the cosine transform. It also plays a role in partial differential equation solutions. Although program D12R5 is again the same as D12R3, you will notice some difference in solutions. The cosine transform of a cosine with a period that is a power of two does not give a transform that is nonzero in a single bin. It has some small values at other frequencies. This is due to our desire to cast the transform into something that calls REALFT, and therefore works on 2^N points rather than the more natural $2^N + 1$. The sample program will prove to you, however, that the transform expressed here is invertible. Notice that, unlike the sine transform, the cosine transform is not self-inverting.

```
      PROGRAM D12R5
C     Driver for routine COSFT
      PARAMETER(EPS=1.0E-3,NP=16,WIDTH=30.0,PI=3.14159)
      DIMENSION DATA(NP),SIZE(NP)
1     WRITE(*,'(1X,A,I2,A)') 'Period of cosine in channels (2-',NP,')'
      READ(*,*) PER
      IF (PER.LE.0.) STOP
      DO 11 I=1,NP
        DATA(I)=COS(2.0*PI*(I-1)/PER)
11    CONTINUE
      CALL COSFT(DATA,NP,+1)
      BIG=-1.0E10
      SMALL=1.0E10
      DO 12 I=1,NP
        IF (DATA(I).LT.SMALL) SMALL=DATA(I)
        IF (DATA(I).GT.BIG) BIG=DATA(I)
12    CONTINUE
      SCAL=WIDTH/(BIG-SMALL)
      DO 13 I=1,NP
        NLIM=SCAL*(DATA(I)-SMALL)+EPS
        WRITE(*,'(1X,I2,F6.2,1X,60A1)') I,DATA(I),('*',J=1,NLIM+1)
13    CONTINUE
      WRITE(*,*) 'press continue ...'
      READ(*,*)
      CALL COSFT(DATA,NP,-1)
      BIG=-1.0E10
      SMALL=1.0E10
      DO 14 I=1,NP
        IF(DATA(I).LT.SMALL) SMALL=DATA(I)
        IF(DATA(I).GT.BIG) BIG=DATA(I)
14    CONTINUE
      SCAL=WIDTH/(BIG-SMALL)
      DO 15 I=1,NP
        NLIM=SCAL*(DATA(I)-SMALL)+EPS
        WRITE(*,'(1X,I4,1X,60A1)') I,('*',J=1,NLIM+1)
15    CONTINUE
      GOTO 1
      END
```

Subroutine CONVLV performs the convolution of a data set with a response function using an FFT. Sample program D12R6 uses two functions that take on only the values 0.0 and 1.0. The data array DATA(I) has sixteen values, and is zero everywhere except between I=6 and I=10 where it is 1.0. The response function RESPNS(I) has nine values and is zero except between I=3 and I=6 where it is 1.0. The expected value of the convolution is determined simply by flipping the response function end-to-end, moving it to the left by the desired shift, and counting how many non-zero channels of RESPNS fall on non-zero channels of DATA. In this way, you should be able to verify the result from the program. The sample program, incidentally, does the calculation by this direct method for the purpose of comparison.

```
      PROGRAM D12R6
C     Driver for routine CONVLV
      PARAMETER(N=16,N2P2=34,M=9,PI=3.14159265)
      DIMENSION DATA(N),RESPNS(M),RESP(N),ANS(N2P2)
      DO 11 I=1,N
```

```
            DATA(I)=0.0
            IF ((I.GE.(N/2-N/8)).AND.(I.LE.(N/2+N/8))) DATA(I)=1.0
11      CONTINUE
        DO 12 I=1,M
            RESPNS(I)=0.0
            IF (I.GT.2 .AND. I.LT.7) RESPNS(I)=1.0
            RESP(I)=RESPNS(I)
12      CONTINUE
        ISIGN=1
        CALL CONVLV(DATA,N,RESP,M,ISIGN,ANS)
C       Compare with a direct convolution
        WRITE(*,'(/1X,T4,A,T13,A,T24,A)') 'I','CONVLV','Expected'
        DO 14 I=1,N
            CMP=0.0
            DO 13 J=1,M/2
                CMP=CMP+DATA(MOD(I-J-1+N,N)+1)*RESPNS(J+1)
                CMP=CMP+DATA(MOD(I+J-1,N)+1)*RESPNS(M-J+1)
13          CONTINUE
            CMP=CMP+DATA(I)*RESPNS(1)
            WRITE(*,'(1X,I3,3X,2F12.6)') I,ANS(I),CMP
14      CONTINUE
        END
```

CORREL calculates the correlation function of two data sets. Sample program D12R7 defines DATA1(I) as an array of 64 values which are all zero except from I=25 to I=39, where they are one. DATA2(I) is defined in the same way. Therefore, the correlation being performed is an autocorrelation. The sample routine compares the result of the calculation as performed by CORREL with that found by a direct calculation. In this case the calculation may be done manually simply by successively shifting DATA2 with respect to DATA1 and counting the number of nonzero channels of the two that overlap.

```
        PROGRAM D12R7
C       Driver for routine CORREL
        PARAMETER(N=64,N2=128,PI=3.1415927)
        DIMENSION DATA1(N),DATA2(N),ANS(N2)
        DO 11 I=1,N
            DATA1(I)=0.0
            IF ((I.GT.(N/2-N/8)).AND.(I.LT.(N/2+N/8))) DATA1(I)=1.0
            DATA2(I)=DATA1(I)
11      CONTINUE
        CALL CORREL(DATA1,DATA2,N,ANS)
C       Calculate directly
        WRITE(*,'(/1X,T4,A,T13,A,T25,A/)') 'n','CORREL','Direct Calc.'
        DO 13 I=0,16
            CMP=0.0
            DO 12 J=1,N
                CMP=CMP+DATA1(MOD(I+J-1,N)+1)*DATA2(J)
12          CONTINUE
            WRITE(*,'(1X,I3,3X,F12.6,F15.6)') I,ANS(I+1),CMP
13      CONTINUE
        END
```

SPCTRM does a spectral estimate of a data set by reading it in as segments, windowing, Fourier transforming, and accumulating the power spectrum. Data segments may or may not be overlapped at the decision of the user. In sample program D12R8 the spectral data is read in from a file called SPCTRL.DAT containing 1200 numbers and included on the *Numerical Recipes Examples Diskette.* It is analyzed first with overlap and then without. The results are tabulated side by side for comparison.

```
      PROGRAM D12R8
C     Driver for routine SPCTRM
      PARAMETER(M=16,M4=4*M)
      DIMENSION P(M),Q(M),W1(M4),W2(M)
      LOGICAL OVRLAP
      OPEN(9,FILE='SPCTRL.DAT',STATUS='OLD')
      K=8
      OVRLAP=.TRUE.
      CALL SPCTRM(P,M,K,OVRLAP,W1,W2)
      REWIND(9)
      K=16
      OVRLAP=.FALSE.
      CALL SPCTRM(Q,M,K,OVRLAP,W1,W2)
      CLOSE(9)
      WRITE(*,*) 'Spectrum of data in file SPCTRL.DAT'
      WRITE(*,'(1X,T14,A,T29,A)') 'Overlapped','Non-Overlapped'
      DO 11 J=1,M
        WRITE(*,'(1X,I4,2F17.6)') J,P(J),Q(J)
11    CONTINUE
      END
```

MEMCOF and EVLMEM are used to perform spectral analysis by the maximum entropy method. MEMCOF finds the coefficients for a model spectrum, the magnitude squared of the inverse of a polynomial series. Sample program D12R9 determines the coefficients for 1000 numbers from the file SPCTRL.DAT and simply prints the results for comparison to the following table:

```
Coefficients for spectral estimation of SPCTRL.DAT
 a[ 1] =     1.261539
 a[ 2] =    -0.007695
 a[ 3] =    -0.646778
 a[ 4] =    -0.280603
 a[ 5] =     0.163693
 a[ 6] =     0.347674
 a[ 7] =     0.111247
 a[ 8] =    -0.337141
 a[ 9] =    -0.358043
 a[10] =     0.378774
    b0 =     0.003511
```

```
      PROGRAM D12R9
C     Driver for routine MEMCOF
      PARAMETER(N=1000,M=10)
      DIMENSION DATA(N),COF(M),WK1(N),WK2(N),WKM(M)
      OPEN(5,FILE='SPCTRL.DAT',STATUS='OLD')
      READ(5,*) (DATA(I),I=1,N)
      CLOSE(5)
      CALL MEMCOF(DATA,N,M,PM,COF,WK1,WK2,WKM)
      WRITE(*,'(/1X,A/)') 'Coeff. for spectral estim. of SPCTRL.DAT'
      DO 11 I=1,M
        WRITE(*,'(1X,A,I2,A,F12.6)') 'a[',I,'] =',COF(I)
11    CONTINUE
```

```
      WRITE(*,'(/1X,A,F12.6/)') 'b0 =',PM
      END
```

EVLMEM uses coefficients from MEMCOF to generate a spectral estimate. The example D12R10 uses the same data from SPCTRL.DAT and prints the spectral estimate. You may compare the result to:

```
Power spectrum estimate of data in SPCTRL.DAT
     f*delta        power
     0.000000     0.026023
     0.031250     0.029266
     0.062500     0.193087
     0.093750     0.139241
     0.125000    29.915518
     0.156250     0.003878
     0.187500     0.000633
     0.218750     0.000334
     0.250000     0.000437
     0.281250     0.001331
     0.312500     0.000780
     0.343750     0.000451
     0.375000     0.000784
     0.406250     0.001381
     0.437500     0.000649
     0.468750     0.000775
     0.500000     0.001716
```

```
      PROGRAM D12R10
C     Driver for routine EVLMEM
      PARAMETER(N=1000,M=10,NFDT=16)
      DIMENSION DATA(N),COF(M),WK1(N),WK2(N),WKM(M)
      OPEN(5,FILE='SPCTRL.DAT',STATUS='OLD')
      READ(5,*) (DATA(I),I=1,N)
      CLOSE(5)
      CALL MEMCOF(DATA,N,M,PM,COF,WK1,WK2,WKM)
      WRITE(*,*) 'Power spectrum estimate of data in SPCTRL.DAT'
      WRITE(*,'(1X,T6,A,T20,A)') 'f*delta','power'
      DO 11 I=0,NFDT
          FDT=0.5*I/NFDT
          WRITE(*,'(1X,2F12.6)') FDT,EVLMEM(FDT,COF,M,PM)
11    CONTINUE
      END
```

Notice that once MEMCOF has determined coefficients, we may evaluate the estimate at any intervals we wish. Notice also that we have built a spectral peak into the noisy data in SPCTRL.DAT.

Linear prediction is carried out by routines PREDIC, MEMCOF, and FIXRTS. MEMCOF produces the linear prediction coefficients from the data set. FIXRTS massages the coefficients so that all roots of the characteristic polynomial fall inside the unit circle of the complex domain, thus insuring stability of the prediction algorithm. Finally, PREDIC predicts future data points based on the modified coefficients. Sample program D12R11 demonstrates the operation of FIXRTS. The coefficients provided in the DATA statement for D(I) are those appropriate to the polynomial $(z-1)^6 = 1$. This equation has six roots on a circle of radius one, centered at $(1.0, 0.0)$ in the complex plane. Some of these lie within the unit circle and some outside. The ones outside are moved by FIXRTS according to $z_i \to 1/z_i^*$. You can easily figure these out by hand and check the results. Also, the sample routine calculates $(z-1)^6$ for each of the adjusted roots, and thereby shows which have been changed and which have not.

```
      PROGRAM D12R11
C     Driver for routine FIXRTS
      PARAMETER(NPOLES=6,NPOL=NPOLES+1)
      DIMENSION D(NPOLES)
      COMPLEX ZCOEF(NPOL),ZEROS(NPOLES),Z
      LOGICAL POLISH
      DATA D/6.0,-15.0,20.0,-15.0,6.0,0.0/
C     Finding roots of (z-1.0)**6=1.0
C     First print roots
      ZCOEF(NPOLES+1)=CMPLX(1.0,0.0)
      DO 11 I=NPOLES,1,-1
        ZCOEF(I)=CMPLX(-D(NPOLES+1-I),0.0)
11    CONTINUE
      POLISH=.TRUE.
      CALL ZROOTS(ZCOEF,NPOLES,ZEROS,POLISH)
      WRITE(*,'(/1X,A)') 'Roots of (z-1.0)^6 = 1.0'
      WRITE(*,'(1X,T20,A,T42,A)') 'Root','(z-1.0)^6'
      DO 12 I=1,NPOLES
        Z=(ZEROS(I)-1.0)**6
        WRITE(*,'(1X,I6,4F12.6)') I,ZEROS(I),Z
12    CONTINUE
C     Now fix them to lie within unit circle
      CALL FIXRTS(D,NPOLES)
C     Check results
      ZCOEF(NPOLES+1)=CMPLX(1.0,0.0)
      DO 13 I=NPOLES,1,-1
        ZCOEF(I)=CMPLX(-D(NPOLES+1-I),0.0)
13    CONTINUE
      CALL ZROOTS(ZCOEF,NPOLES,ZEROS,POLISH)
      WRITE(*,'(/1X,A)') 'Roots reflected in unit circle'
      WRITE(*,'(1X,T20,A,T42,A)') 'Root','(z-1.0)^6'
      DO 14 I=1,NPOLES
        Z=(ZEROS(I)-1.0)**6
        WRITE(*,'(1X,I6,4F12.6)') I,ZEROS(I),Z
14    CONTINUE
      END
```

PREDIC carries out the job of performing the prediction. The function chosen for investigation in sample program D12R12 is

$$F(\mathtt{N}) = \exp(-\mathtt{N}/\mathtt{NPTS})\sin(2\pi\mathtt{N}/50) + \exp(-2\mathtt{N}/\mathtt{NPTS})\sin(2.2\pi\mathtt{N}/50)$$

the sum of two sine waves of similar period and exponentially decaying amplitudes. On the basis of 500 data points, and working with coefficients representing ten poles, the routine predicts 20 future points. The quality of this prediction may be judged by comparing these 20 points with the evaluations of $F(\mathtt{N})$ that are provided.

```
      PROGRAM D12R12
C     Driver for routine PREDIC
      PARAMETER(NPTS=500,NPOLES=10,NFUT=20,PI=3.1415926)
      DIMENSION DATA(NPTS),D(NPOLES),WK1(NPTS),
```

```
     *         WK2(NPTS),WKM(NPOLES),FUTURE(NFUT)
      F(N)=EXP(-1.0*N/NPTS)*SIN(2.0*PI*N/50.0)
     *         +EXP(-2.0*N/NPTS)*SIN(2.2*PI*N/50.0)
      DO 11 I=1,NPTS
        DATA(I)=F(I)
11    CONTINUE
      CALL MEMCOF(DATA,NPTS,NPOLES,DUM,D,WK1,WK2,WKM)
      CALL FIXRTS(D,NPOLES)
      CALL PREDIC(DATA,NPTS,D,NPOLES,FUTURE,NFUT)
      WRITE(*,'(6X,A,T13,A,T25,A)') 'I','Actual','PREDIC'
      DO 12 I=1,NFUT
        WRITE(*,'(1X,I6,2F12.6)') I,F(I+NPTS),FUTURE(I)
12    CONTINUE
      END
```

FOURN is a routine for performing N-dimensional Fourier transforms. We have used it in sample program D12R13 to transform a 3-dimensional complex data array of dimensions $4 \times 8 \times 16$. The function analyzed is not that easy to visualize, but it is very easy to calculate. The test conducted here is to perform a 3-dimensional transform and inverse transform in succession, and to compare the result with the original array. Ratios are provided for convenience.

```
      PROGRAM D12R13
C     Driver for routine FOURN
      PARAMETER(NDIM=3,NDAT=1024)
      DIMENSION NN(NDIM),DATA(NDAT)
      DO 11 I=1,NDIM
        NN(I)=2*(2**I)
11    CONTINUE
      DO 14 I=1,NN(3)
        DO 13 J=1,NN(2)
          DO 12 K=1,NN(1)
            L=K+(J-1)*NN(1)+(I-1)*NN(2)*NN(1)
            LL=2*L-1
            DATA(LL)=FLOAT(LL)
            DATA(LL+1)=FLOAT(LL+1)
12        CONTINUE
13      CONTINUE
14    CONTINUE
      ISIGN=+1
      CALL FOURN(DATA,NN,NDIM,ISIGN)
      ISIGN=-1
      WRITE(*,'(1X,A)') 'Double 3-dimensional Transform'
      WRITE(*,'(/1X,T10,A,T35,A,T63,A)') 'Double Transf.',
     *         'Original Data','Ratio'
      WRITE(*,'(1X,T8,A,T20,A,T33,A,T45,A,T57,A,T69,A/)')
     *         'Real','Imag.','Real','Imag.','Real','Imag.'
      CALL FOURN(DATA,NN,NDIM,ISIGN)
      DO 15 I=1,4
        J=2*I
        K=2*J
        L=K+(J-1)*NN(1)+(I-1)*NN(2)*NN(1)
        LL=2*L-1
        WRITE(*,'(1X,6F12.2)') DATA(LL),DATA(LL+1),FLOAT(LL),FLOAT(LL+1),
     *           DATA(LL)/LL,DATA(LL+1)/(LL+1)
15    CONTINUE
```

```
      WRITE(*,'(/1X,A,I4)') 'The product of transform lengths is:',
     *          NN(1)*NN(2)*NN(3)
      END
```

Chapter 13: Statistical Description of Data

Chapter 13 of Numerical Recipes covers the subject of descriptive statistics, the representation of data in terms of its statistical properties, and the use of such properties to compare data sets. There are three subroutines that characterize data sets. MOMENT *returns the average, average deviation, standard deviation, variance, skewness, and kurtosis of a data array.* MDIAN1 *and* MDIAN2 *both find the median of an array. The former also sorts the array.*

Most of the remaining subroutines compare data sets. TTEST *compares the means of two data sets having the same variance;* TUTEST *does the same for two sets having different variance; and* TPTEST *does it for paired samples, correcting for covariance.* FTEST *is a test of whether two data arrays have significantly different variance. The question of whether two distributions are different is treated by four subroutines (pertaining to whether the data is binned or continuous, and whether data is compared to a model distribution or to other data). Specifically,*

1. CHSONE *compares binned data to a model distribution.*
2. CHSTWO *compares two binned data sets.*
3. KSONE *compares the cumulative distribution function of an unbinned data set to a given function.*
4. KSTWO *compares the cumulative distribution functions of two unbinned data sets.*

The next set of subroutines tests for associations between nominal variables. CNTAB1 *and* CNTAB2 *both check for associations in a two-dimensional contingency table, the first calculating on the basis of* χ^2*, and the second by evaluating entropies. Linear correlation is represented by Pearson's* r*, or the linear correlation coefficient, which is calculated with routine* PEARSN. *Alternatively, the data can be investigated with a nonparametric or rank correlation, using* SPEAR *to find Spearman's rank correlation* r_s*. Kendall's* τ *uses rank ordering of ordinal data to test for monotonic correlations.* KENDL1 *does this for two data arrays of the same size, while* KENDL2 *applies it to contingency tables.*

One final routine SMOOFT *makes no attempt to describe or compare data statistically. It seeks, instead, to smooth out the statistical fluctuations, usually for the purpose of visual presentation.*

⋆ ⋆ ⋆ ⋆

Subroutine MOMENT calculates successive moments of a given distribution of data. The example program D13R1 creates an unusual distribution, one that has a sinusoidal distribution of values (over a half-period of the sine, so the distribution is a symmetrical peak). We have worked out the moments of such a distribution theoretically and recorded them in the program for comparison. The data is discrete and will only approximate these values.

```
      PROGRAM D13R1
C     Driver for routine MOMENT
      PARAMETER(PI=3.14159265,NPTS=10000,NBIN=100,NDAT=NPTS+NBIN)
      DIMENSION DATA(NDAT)
      I=1
      DO 12 J=1,NBIN
        X=PI*J/NBIN
        NLIM=NINT(SIN(X)*PI/2.0*NPTS/NBIN)
        DO 11 K=1,NLIM
          DATA(I)=X
          I=I+1
11      CONTINUE
12    CONTINUE
      WRITE(*,'(1X,A/)') 'Moments of a sinusoidal distribution'
      CALL MOMENT(DATA,NPTS,AVE,ADEV,SDEV,VAR,SKEW,CURT)
      WRITE(*,'(1X,T29,A,T42,A/)') 'Calculated','Expected'
      WRITE(*,'(1X,A,T25,2F12.4)') 'Mean :',AVE,PI/2.0
      WRITE(*,'(1X,A,T25,2F12.4)') 'Average Deviation :',ADEV,0.570796
      WRITE(*,'(1X,A,T25,2F12.4)') 'Standard Deviation :',SDEV,0.683667
      WRITE(*,'(1X,A,T25,2F12.4)') 'Variance :',VAR,0.467401
      WRITE(*,'(1X,A,T25,2F12.4)') 'Skewness :',SKEW,0.0
      WRITE(*,'(1X,A,T25,2F12.4)') 'Kurtosis :',CURT,-0.806249
      END
```

MDIAN1 and MDIAN2 both find the median of a distribution. In programs D13R2 and D13R3 we allow this distribution to be Gaussian, as produced by routine GASDEV. This distribution should have a mean of zero and variance of one. MDIAN1 also sorts the data, so D13R2 prints the sorted data to show that it is done properly. Example D13R3 has nothing to show from MDIAN2 but the median itself, and it is checked by comparing to the result from MDIAN1.

```
      PROGRAM D13R2
C     Driver for routine MDIAN1
      PARAMETER(NPTS=50)
      DIMENSION DATA(NPTS)
      IDUM=-5
      DO 11 I=1,NPTS
        DATA(I)=GASDEV(IDUM)
11    CONTINUE
      CALL MDIAN1(DATA,NPTS,XMED)
      WRITE(*,'(1X,A/)') 'Gaussian distrib., zero mean, unit variance'
      WRITE(*,'(1X,A,F10.6/)') 'Median of data set is',XMED
      WRITE(*,'(1X,A/,(5F12.6))') 'Sorted data',
     *          (DATA(I),I=1,50)
      END

      PROGRAM D13R3
C     Driver for routine MDIAN2
      PARAMETER(NPTS=50)
```

```
      DIMENSION DATA(NPTS)
      IDUM=-5
      DO 11 I=1,NPTS
        DATA(I)=GASDEV(IDUM)
11    CONTINUE
      CALL MDIAN2(DATA,NPTS,XMED)
      WRITE(*,'(1X,A/)') 'Gaussian distrib., zero mean, unit variance'
      WRITE(*,'(1X,A,F12.6)') 'Median according to MDIAN2 is',XMED
      CALL MDIAN1(DATA,NPTS,XMED)
      WRITE(*,'(1X,A,F12.6/)') 'Median according to MDIAN1 is',XMED
      END
```

Student's t-test is a test of two data sets for significantly different means. It is applied by D13R4 to two Gaussian data sets DATA1 and DATA2 that are generated by GASDEV. DATA2 is originally given an artificial shift of its mean to the right of that of DATA1, by NSHFT/2 units of EPS. Then DATA1 is successively shifted NSHFT times to the right by EPS and compared to DATA2 by TTEST. At about step NSHFT/2, the two distributions should superpose and indicate populations with the same mean. Notice that the two populations have the same variance (i.e. 1.0), as required by TTEST.

```
      PROGRAM D13R4
C     Driver for routine TTEST
      PARAMETER(NPTS=1024, MPTS=512, EPS=0.02,NSHFT=10)
      DIMENSION DATA1(NPTS),DATA2(MPTS)
C     Generate Gaussian distributed data
      IDUM=-5
      DO 11 I=1,NPTS
        DATA1(I)=GASDEV(IDUM)
11    CONTINUE
      IDUM=-11
      DO 12 I=1,MPTS
        DATA2(I)=(NSHFT/2.0)*EPS+GASDEV(IDUM)
12    CONTINUE
      WRITE(*,'(/1X,T4,A,T18,A,T25,A)') 'Shift','T','Probability'
      DO 14 I=1,NSHFT+1
        CALL TTEST(DATA1,NPTS,DATA2,MPTS,T,PROB)
        SHIFT=(I-1)*EPS
        WRITE(*,'(1X,F6.2,2F12.2)') SHIFT,T,PROB
        DO 13 J=1,NPTS
          DATA1(J)=DATA1(J)+EPS
13      CONTINUE
14    CONTINUE
      END
```

AVEVAR is an auxiliary routine for TTEST. It finds the average and variance of a data set. Sample program D13R5 generates a series of Gaussian distributions for I=1,...,11, and gives each a shift of (I − 1)EPS and a variance of I^2. This progression allows you easily to check the operation of AVEVAR "by eye".

```
      PROGRAM D13R5
C     Driver for routine AVEVAR
      PARAMETER(NPTS=1000, EPS=0.1)
      DIMENSION DATA(NPTS)
C     Generate Gaussian distributed data
      IDUM=-5
```

```
      WRITE(*,'(1X,T4,A,T14,A,T26,A)') 'Shift','Average','Variance'
      DO 12 I=1,11
          SHIFT=(I-1)*EPS
          DO 11 J=1,NPTS
              DATA(J)=SHIFT+I*GASDEV(IDUM)
11        CONTINUE
          CALL AVEVAR(DATA,NPTS,AVE,VAR)
          WRITE(*,'(1X,F6.2,2F12.2)') SHIFT,AVE,VAR
12    CONTINUE
      END
```

TUTEST also does Student's t-test, but applies to the comparison of means of two distributions with different variance. The example D13R6A employs the comparison used on TTEST but gives the two distributions DATA1 and DATA2 variances of 1.0 and 4.0 respectively.

```
      PROGRAM D13R6A
C     Driver for routine TUTEST
      PARAMETER(NPTS=5000,MPTS=1000,EPS=0.02,VAR1=1.0,
     *        VAR2=4.0,NSHFT=10)
      DIMENSION DATA1(NPTS),DATA2(MPTS)
C     Generate two Gaussian distributions of different variance
      IDUM=-51773
      FCTR1=SQRT(VAR1)
      DO 11 I=1,NPTS
          DATA1(I)=FCTR1*GASDEV(IDUM)
11    CONTINUE
      FCTR2=SQRT(VAR2)
      DO 12 I=1,MPTS
          DATA2(I)=(NSHFT/2.0)*EPS+FCTR2*GASDEV(IDUM)
12    CONTINUE
      WRITE(*,'(1X,A,F6.2)') 'Distribution #1 : variance = ',VAR1
      WRITE(*,'(1X,A,F6.2/)') 'Distribution #2 : variance = ',VAR2
      WRITE(*,'(1X,T4,A,T18,A,T25,A)') 'Shift','T','Probability'
      DO 14 I=1,NSHFT+1
          CALL TUTEST(DATA1,NPTS,DATA2,MPTS,T,PROB)
          SHIFT=(I-1)*EPS
          WRITE(*,'(1X,F6.2,2F12.2)') SHIFT,T,PROB
          DO 13 J=1,NPTS
              DATA1(J)=DATA1(J)+EPS
13        CONTINUE
14    CONTINUE
      END
```

TPTEST goes a step further, and compares two distributions not only having different variances, but also perhaps having point by point correlations. The example D13R6B creates two situations, one with correlated and one with uncorrelated distributions. It does this by way of three data sets. DATA1 is a simple Gaussian distribution of zero mean and unit variance. DATA2 is DATA1 plus some additional Gaussian fluctuations of smaller amplitude. DATA3 is similar to DATA2 but generated with independent calls to GASDEV so that its fluctuations ought not to have any correlation with those of DATA1. DATA1 is then given an offset with respect to the others and they are successively shifted as in previous routines. At each step of the shift TPTEST was applied. Our results are given below:

	Correlated:		Uncorrelated:	
Shift	T	Probability	T	Probability
.01	2.9264	0.0036	0.6028	0.5469
.02	2.1948	0.0286	0.4521	0.6514
.03	1.4632	0.1440	0.3014	0.7632
.04	0.7316	0.4647	0.1507	0.8802
.05	0.0000	1.0000	0.0000	1.0000
.06	-0.7316	0.4647	-0.1507	0.8802
.07	-1.4632	0.1440	-0.3014	0.7632
.08	-2.1948	0.0286	-0.4521	0.6514
.09	-2.9264	0.0036	-0.6028	0.5469
.10	-3.6580	0.0003	-0.7536	0.4514
.11	-4.3896	0.0000	-0.9043	0.3663

```
      PROGRAM D13R6B
C     Driver for routine TPTEST
C     Compare two correlated distributions vs. two
C     uncorrelated distributions
      PARAMETER(NPTS=500,EPS=0.01,NSHFT=10,ANOISE=0.3)
      DIMENSION DATA1(NPTS),DATA2(NPTS),DATA3(NPTS)
      IDUM=-5
      WRITE(*,'(1X,T18,A,T46,A)') 'Correlated:','Uncorrelated:'
      WRITE(*,'(1X,T4,A,T18,A,T25,A,T46,A,T53,A)')
     *       'Shift','T','Probability','T','Probability'
      OFFSET=(NSHFT/2)*EPS
      DO 11 J=1,NPTS
        GAUSS=GASDEV(IDUM)
        DATA1(J)=GAUSS
        DATA2(J)=GAUSS+ANOISE*GASDEV(IDUM)
        DATA3(J)=GASDEV(IDUM)+ANOISE*GASDEV(IDUM)
11    CONTINUE
      CALL AVEVAR(DATA1,NPTS,AVE1,VAR1)
      CALL AVEVAR(DATA2,NPTS,AVE2,VAR2)
      CALL AVEVAR(DATA3,NPTS,AVE3,VAR3)
      DO 12 J=1,NPTS
        DATA1(J)=DATA1(J)-AVE1+OFFSET
        DATA2(J)=DATA2(J)-AVE2
        DATA3(J)=DATA3(J)-AVE3
12    CONTINUE
      DO 14 I=1,NSHFT+1
        SHIFT=I*EPS
        DO 13 J=1,NPTS
          DATA2(J)=DATA2(J)+EPS
          DATA3(J)=DATA3(J)+EPS
13      CONTINUE
        CALL TPTEST(DATA1,DATA2,NPTS,T1,PROB1)
        CALL TPTEST(DATA1,DATA3,NPTS,T2,PROB2)
        WRITE(*,'(1X,F6.2,2X,2F12.4,4X,2F12.4)')
     *         SHIFT,T1,PROB1,T2,PROB2
14    CONTINUE
      END
```

The F-test (subroutine FTEST) is a test for differing variances between two distributions. For demonstration purposes, sample program D13R7 generates two distributions DATA1 and DATA2 having Gaussian distributions of unit variance. The values of a third array DATA3 are then set by multiplying DATA2 by a series of values

FACTR which takes its variance from 1.0 to 1.1 in ten equal steps. The effect of this on the F-test can be evaluated from the probabilities PROB.

```
      PROGRAM D13R7
C     Driver for routine FTEST
      PARAMETER(NPTS=1000,MPTS=500,EPS=0.01,NVAL=10)
      DIMENSION DATA1(NPTS),DATA2(MPTS),DATA3(MPTS)
C     Generate two Gaussian distributions with
C     different variances
      IDUM=-13
      DO 11 J=1,NPTS
        DATA1(J)=GASDEV(IDUM)
11    CONTINUE
      DO 12 J=1,MPTS
        DATA2(J)=GASDEV(IDUM)
12    CONTINUE
      WRITE(*,'(1X,T5,A,F5.2)') 'Variance 1 = ',1.0
      WRITE(*,'(1X,T5,A,T21,A,T30,A)')
     *          'Variance 2','Ratio','Probability'
      DO 14 I=1,NVAL+1
        VAR=1.0+(I-1)*EPS
        FACTOR=SQRT(VAR)
        DO 13 J=1,MPTS
          DATA3(J)=FACTOR*DATA2(J)
13      CONTINUE
        CALL FTEST(DATA1,NPTS,DATA3,MPTS,F,PROB)
        WRITE(*,'(1X,F11.4,2X,2F12.4)') VAR,F,PROB
14    CONTINUE
      END
```

CHSONE and CHSTWO compare two distributions on the basis of a χ^2 test to see if they are different. CHSONE, specifically, compares a data distribution to an expected distribution. Sample program D13R8 generates an exponential distribution BINS(I) of data using routine EXPDEV. It then creates an array EBINS(I) which is the expected result (a smooth exponential decay in the absence of statistical fluctuations). EBINS and BINS are compared by CHSONE to give χ^2 and a probability that they represent the same distribution.

```
      PROGRAM D13R8
C     Driver for routine CHSONE
      PARAMETER(NBINS=10,NPTS=2000)
      DIMENSION BINS(NBINS),EBINS(NBINS)
      IDUM=-15
      DO 11 J=1,NBINS
        BINS(J)=0.0
11    CONTINUE
      DO 12 I=1,NPTS
        X=EXPDEV(IDUM)
        IBIN=X*NBINS/3.0+1
        IF(IBIN.LE.NBINS) BINS(IBIN)=BINS(IBIN)+1.0
12    CONTINUE
      DO 13 I=1,NBINS
        EBINS(I)=3.0*NPTS/NBINS*EXP(-3.0*(I-0.5)/NBINS)
13    CONTINUE
      CALL CHSONE(BINS,EBINS,NBINS,-1,DF,CHSQ,PROB)
      WRITE(*,'(1X,T10,A,T25,A)') 'Expected','Observed'
      DO 14 I=1,NBINS
```

```
        WRITE(*,'(1X,2F15.2)') EBINS(I),BINS(I)
14    CONTINUE
      WRITE(*,'(/1X,T9,A,E12.4)') 'Chi-squared:',CHSQ
      WRITE(*,'(1X,T9,A,E12.4)') 'Probability:',PROB
      END
```

CHSTWO compares two binned distributions BINS1 and BINS2, again using a χ^2 test. Sample program D13R9 prepares these distributions both in the same way. Each is composed of 2000 random numbers, drawn from an exponential deviate, and placed into 10 bins. The two data sets are then analyzed by CHSTWO to calculate χ^2 and probability PROB.

```
      PROGRAM D13R9
C     Driver for routine CHSTWO
      PARAMETER(NBINS=10,NPTS=2000)
      DIMENSION BINS1(NBINS),BINS2(NBINS)
      IDUM=-17
      DO 11 J=1,NBINS
        BINS1(J)=0.0
        BINS2(J)=0.0
11    CONTINUE
      DO 12 I=1,NPTS
        X=EXPDEV(IDUM)
        IBIN=X*NBINS/3.0+1
        IF(IBIN.LE.NBINS) BINS1(IBIN)=BINS1(IBIN)+1.0
        X=EXPDEV(IDUM)
        IBIN=X*NBINS/3.0+1
        IF(IBIN.LE.NBINS) BINS2(IBIN)=BINS2(IBIN)+1.0
12    CONTINUE
      CALL CHSTWO(BINS1,BINS2,NBINS,-1,DF,CHSQ,PROB)
      WRITE(*,'(1X,T10,A,T25,A)') 'Dataset 1','Dataset 2'
      DO 13 I=1,NBINS
        WRITE(*,'(1X,2F15.2)') BINS1(I),BINS2(I)
13    CONTINUE
      WRITE(*,'(/1X,T10,A,E12.4)') 'Chi-squared:',CHSQ
      WRITE(*,'(1X,T10,A,E12.4)') 'Probability:',PROB
      END
```

The Kolmogorov-Smirnov test used in KSONE and KSTWO applies to unbinned distributions with a single independent variable. KSONE uses the K-S criterion to compare a single data set to an expected distribution, and KSTWO uses it to compare two data sets. Sample program D13R10 creates data sets with Gaussian distributions and with stepwise increasing variance, and compares their cumulative distribution function to the expected result for a Gaussian distribution of unit variance. This result is the error function and is generated by routine ERF. Increasing variance in the test destribution should reduce the likelihood that it was drawn from the same distribution represented by the comparison function.

```
      PROGRAM D13R10
C     Driver for routine KSONE
      PARAMETER(NPTS=1000,EPS=0.1)
      DIMENSION DATA(NPTS)
      EXTERNAL FUNC
      IDUM=-5
      WRITE(*,'(/1X,T5,A,T24,A,T44,A/)')
     *          'Variance Ratio','K-S Statistic','Probability'
```

```
      DO 12 I=1,11
          VAR=1.0+(I-1)*EPS
          FACTR=SQRT(VAR)
          DO 11 J=1,NPTS
              DATA(J)=FACTR*ABS(GASDEV(IDUM))
11        CONTINUE
          CALL KSONE(DATA,NPTS,FUNC,D,PROB)
          WRITE(*,'(1X,F14.6,F18.6,E20.4)') VAR,D,PROB
12    CONTINUE
      END
      FUNCTION FUNC(X)
      Y=X/SQRT(2.0)
      FUNC=ERF(Y)
      END
```

KSTWO compares the cumulative distribution functions of two unbinned data sets, DATA1 and DATA2. In sample program D13R11, they are both Gaussian distributions, but DATA2 is given a stepwise increase of variance. In other respects, D13R11 is like D13R10.

```
      PROGRAM D13R11
C     Driver for routine KSTWO
      PARAMETER(N1=2000,N2=1000,EPS=0.1)
      DIMENSION DATA1(N1),DATA2(N2)
      IDUM=-1357
      DO 11 J=1,N1
          DATA1(J)=GASDEV(IDUM)
11    CONTINUE
      WRITE(*,'(/1X,T6,A,T26,A,T46,A/)')
     *        'Variance Ratio','K-S Statistic','Probability'
      IDUM=-2468
      DO 13 I=1,11
          VAR=1.0+(I-1)*EPS
          FACTR=SQRT(VAR)
          DO 12 J=1,N2
              DATA2(J)=FACTR*GASDEV(IDUM)
12        CONTINUE
          CALL KSTWO(DATA1,N1,DATA2,N2,D,PROB)
          WRITE(*,'(1X,F15.6,F19.6,E20.4)') VAR,D,PROB
13    CONTINUE
      END
```

PROBKS is an auxiliary routine for KSONE and KSTWO which calculates the function $Q_{ks}(\lambda)$ used to evaluate the probability that the two distributions being compared are the same. There is no independent means of producing this function, so in sample program D13R12 we have chosen simply to graph it. Our output is reproduced below.

```
      PROGRAM D13R12
C     Driver for routine PROBKS
      CHARACTER TEXT(50)*1
      WRITE(*,*) 'Probability func. for Kolmogorov-Smirnov statistic'
      WRITE(*,'(/1X,T3,A,T15,A,T27,A)') 'Lambda:','Value:','Graph:'
      NPTS=20
      EPS=0.1
      SCALE=40.0
```

```
      DO 12 I=1,NPTS
          ALAM=I*EPS
          VALUE=PROBKS(ALAM)
          TEXT(1)='*'
          DO 11 J=1,50
              IF(J.LE.NINT(SCALE*VALUE)) THEN
                  TEXT(J)='*'
              ELSE
                  TEXT(J)=' '
              ENDIF
11        CONTINUE
          WRITE(*,'(1X,F9.6,F12.6,4X,50A1)') ALAM,VALUE,
     *            (TEXT(J),J=1,50)
12    CONTINUE
      END
```

```
Probability func. for Kolmogorov-Smirnov statistic
 Lambda:     Value:      Graph:
 0.100000    1.000000    ****************************************
 0.200000    1.000000    ****************************************
 0.300000    0.999991    ****************************************
 0.400000    0.997192    ****************************************
 0.500000    0.963945    ***************************************
 0.600000    0.864283    ***********************************
 0.700000    0.711235    ****************************
 0.800000    0.544142    **********************
 0.900000    0.392731    ****************
 1.000000    0.270000    ***********
 1.100000    0.177718    *******
 1.200000    0.112250    ****
 1.300000    0.068092    ***
 1.400000    0.039682    **
 1.500000    0.022218    *
 1.600000    0.011952
 1.700000    0.006177
 1.800000    0.003068
 1.900000    0.001464
 2.000000    0.000671
```

Subroutine CNTAB1 analyzes a two-dimensional contingency table and returns several parameters describing any association between its nominal variables. Sample program D13R3 supplies a table from a file TABLE.DAT which is listed in the Appendix to this chapter. The table shows the rate of certain accidents, tabulated on a monthly basis. These data are listed, as well as their statistical properties, by D13R3. We found the results to be:

```
Chi-squared                     5026.30
Degrees of Freedom                88.00
Probability                       .0000
Cramer-V                          .0772
Contingency Coeff.                .2134
```

```
      PROGRAM D13R13
C     Driver for routine CNTAB1
C     Contingency table in file TABLE.DAT
      PARAMETER(NDAT=9,NMON=12)
```

```
      DIMENSION NMBR(NDAT,NMON)
      CHARACTER FATE(NDAT)*15,MON(NMON)*5,TEXT*64
      OPEN(5,FILE='TABLE.DAT',STATUS='OLD')
      READ(5,*)
      READ(5,'(A)') TEXT
      READ(5,'(15X,12A5/)') (MON(I),I=1,12)
      DO 11 I=1,NDAT
        READ(5,'(A15,12I5)') FATE(I),(NMBR(I,J),J=1,12)
11    CONTINUE
      CLOSE(5)
      WRITE(*,'(/1X,A/)') TEXT
      WRITE(*,'(1X,15X,12A5)') (MON(I),I=1,12)
      DO 12 I=1,NDAT
        WRITE(*,'(1X,A,12I5)') FATE(I),(NMBR(I,J),J=1,12)
12    CONTINUE
      CALL CNTAB1(NMBR,NDAT,NMON,CHISQ,DF,PROB,CRAMRV,CCC)
      WRITE(*,'(/1X,A,T20,F20.2)') 'Chi-squared',CHISQ
      WRITE(*,'(1X,A,T20,F20.2)') 'Degrees of Freedom',DF
      WRITE(*,'(1X,A,T20,F20.4)') 'Probability',PROB
      WRITE(*,'(1X,A,T20,F20.4)') 'Cramer-V',CRAMRV
      WRITE(*,'(1X,A,T20,F20.4)') 'Contingency Coeff.',CCC
      END
```

The test looks for any association between accidents and the months in which they occur. `TABLE.DAT` clearly shows some. Drownings, for example, happen mostly in the summer. `CNTAB2` carries out a similar analysis on `TABLE.DAT` but measures associations on the basis of entropy. Sample program `D13R14` prints out the following entropies for the table:

```
Entropy of Table              4.0368
Entropy of x-distribution     1.5781
Entropy of y-distribution     2.4820
Entropy of y given x          2.4588
Entropy of x given y          1.5548
Dependency of y on x           .0094
Dependency of x on y           .0147
Symmetrical dependency         .0114
```

```
      PROGRAM D13R14
C     Driver for routine CNTAB2
C     Contingency table in file TABLE.DAT
      PARAMETER(NI=9,NMON=12)
      DIMENSION NMBR(NI,NMON)
      CHARACTER FATE(NI)*15,MON(NMON)*5,TEXT*64
      OPEN(5,FILE='TABLE.DAT',STATUS='OLD')
      READ(5,*)
      READ(5,'(A)') TEXT
      READ(5,'(15X,12A5/)') (MON(I),I=1,12)
      DO 11 I=1,NI
        READ(5,'(A15,12I5)') FATE(I),(NMBR(I,J),J=1,12)
11    CONTINUE
      CLOSE(5)
      WRITE(*,'(/1X,A/)') TEXT
      WRITE(*,'(1X,15X,12A5)') (MON(I),I=1,12)
      DO 12 I=1,NI
        WRITE(*,'(1X,A,12I5)') FATE(I),(NMBR(I,J),J=1,12)
12    CONTINUE
```

```
      CALL CNTAB2(NMBR,NI,NMON,H,HX,HY,HYGX,HXGY,UYGX,UXGY,UXY)
      WRITE(*,'(/1X,A,T30,F10.4)') 'Entropy of Table',H
      WRITE(*,'(1X,A,T30,F10.4)') 'Entropy of x-distribution',HX
      WRITE(*,'(1X,A,T30,F10.4)') 'Entropy of y-distribution',HY
      WRITE(*,'(1X,A,T30,F10.4)') 'Entropy of y given x',HYGX
      WRITE(*,'(1X,A,T30,F10.4)') 'Entropy of x given y',HXGY
      WRITE(*,'(1X,A,T30,F10.4)') 'Dependency of y on x',UYGX
      WRITE(*,'(1X,A,T30,F10.4)') 'Dependency of x on y',UXGY
      WRITE(*,'(1X,A,T30,F10.4/)') 'Symmetrical dependency',UXY
      END
```

The dependencies of x on y and y on x indicate the degree to which the type of accident can be predicted by knowing the month, or vice-versa.

PEARSN makes an examination of two ordinal or continuous variables to find linear correlations. It returns a linear correlation coefficient R, a probability of correlation PROB, and Fisher's z. Sample program D13R15 sets up data pairs in arrays DOSE and SPORE which show hypothetical data for the spore count from plants exposed to various levels of γ-rays. The results of applying PEARSN to this data set are compared with the correct results by the program.

```
      PROGRAM D13R15
C     Driver for routine PEARSN
      DIMENSION DOSE(10),SPORE(10)
      DATA DOSE/56.1,64.1,70.0,66.6,82.,91.3,90.,99.7,115.3,110./
      DATA SPORE/0.11,0.4,0.37,0.48,0.75,0.66,0.71,1.2,1.01,0.95/
      WRITE(*,'(1X,A)')
     *        'Effect of Gamma Rays on Man-in-the-Moon Marigolds'
      WRITE(*,'(1X,A,T29,A)') 'Count Rate (cpm)','Pollen Index'
      DO 11 I=1,10
         WRITE(*,'(1X,F10.2,F25.2)') DOSE(I),SPORE(I)
11    CONTINUE
      CALL PEARSN(DOSE,SPORE,10,R,PROB,Z)
      WRITE(*,'(/1X,T24,A,T38,A)') 'PEARSN','Expected'
      WRITE(*,'(1X,A,T18,2E15.6)') 'Corr. Coeff.',R,0.906959
      WRITE(*,'(1X,A,T18,2E15.6)') 'Probability',PROB,0.292650E-3
      WRITE(*,'(1X,A,T18,2E15.6/)') 'Fisher''s Z',Z,1.51011
      END
```

Rank order correlation may be done with SPEAR to compare two distributions DATA1 and DATA2 for correlation. Correlations are reported both in terms of D, the sum-squared difference in ranks, and RS, Spearman's rank correlation parameter. Sample program D13R16 applies the calculation to the data in table TABLE2.DAT (see Appendix) which shows the solar flux incident on various cities during different months of the year. It then checks for correlations between columns of the table, considering each column as a separate data set. In this fashion it looks for correlations between the July solar flux and that of other months. The probability of such correlations are shown by PROBD and PROBRS. Our results are:

Correlation of sampled U.S. solar radiation (July with other months)

Month	D	St. Dev.	PROBD	Spearman R	PROBRS
jul	.00	-4.358899	.000013	.993965	.000000
aug	122.00	-3.958458	.000075	.901959	.000000
sep	218.00	-3.643896	.000269	.832831	.000005
oct	384.00	-3.098495	.001945	.704372	.000526
nov	390.50	-3.077642	.002086	.701205	.000572

```
dec         622.00   -2.318075     .020445          .526751      .017022
jan         644.50   -2.244251     .024816          .509796      .021662
feb         483.50   -2.772503     .005563          .631122      .002844
mar         497.00   -2.728208     .006368          .620949      .003480
apr         405.50   -3.027925     .002462          .688158      .000796
may         264.00   -3.492371     .000479          .794870      .000028
jun         121.50   -3.960099     .000075          .902336      .000000
```

```
      PROGRAM D13R16
C     Driver for routine SPEAR
      PARAMETER(NDAT=20,NMON=12)
      DIMENSION DATA1(NDAT),DATA2(NDAT),RAYS(NDAT,NMON)
      DIMENSION WKSP1(NDAT),WKSP2(NDAT),AVE(NDAT),ZLAT(NDAT)
      CHARACTER CITY(NDAT)*15,MON(NMON)*4,TEXT*64
      OPEN(5,FILE='TABLE2.DAT',STATUS='OLD')
      READ(5,*)
      READ(5,'(A)') TEXT
      READ(5,'(15X,12A4/)') (MON(I),I=1,12)
      DO 11 I=1,NDAT
        READ(5,'(A15,12F4.0,F6.0,F6.1)')
     *          CITY(I),(RAYS(I,J),J=1,12),AVE(I),ZLAT(I)
11    CONTINUE
      CLOSE(5)
      WRITE(*,*) TEXT
      WRITE(*,'(1X,15X,12A4)') (MON(I),I=1,12)
      DO 12 I=1,NDAT
        WRITE(*,'(1X,A,12I4,I6,F6.1)') CITY(I),
     *          (NINT(RAYS(I,J)),J=1,12)
12    CONTINUE
C     Check temperature correlations between different months
      WRITE(*,'(/1X,A)')
     *       'Are sunny summer places also sunny winter places?'
      WRITE(*,'(1X,2A)') 'Check correlation of sampled U.S. solar ',
     * 'radiation (july with other months)'
      WRITE(*,'(/1X,A,T16,A,T23,A,T37,A,T49,A,T63,A/)')
     *       'Month','D','St. Dev.','PROBD',
     *       'Spearman R','PROBRS'
      DO 13 I=1,NDAT
        DATA1(I)=RAYS(I,1)
13    CONTINUE
      DO 15 J=1,12
        DO 14 I=1,NDAT
          DATA2(I)=RAYS(I,J)
14      CONTINUE
        CALL SPEAR(DATA1,DATA2,NDAT,WKSP1,WKSP2,D,ZD,PROBD,RS,PROBRS)
        WRITE(*,'(1X,A,F13.2,2F12.6,3X,2F12.6)')
     *          MON(J),D,ZD,PROBD,RS,PROBRS
15    CONTINUE
      END
```

CRANK is an auxiliary routine for SPEAR and is used in conjunction with SORT2. The latter sorts an array, and CRANK then assigns ranks to each data entry, including the midranking of ties. Sample program D13R17 uses the solar flux data of TABLE2.DAT (see Appendix) to illustrate. Each column of the solar flux table is replaced by the rank order of its entries. You can check the rank order chart against the chart of original values to verify the ordering.

```
      PROGRAM D13R17
C     Driver for routine CRANK
      PARAMETER(NDAT=20,NMON=12)
      DIMENSION DATA(NDAT),RAYS(NDAT,NMON)
      DIMENSION ORDER(NDAT),AVE(NDAT),ZLAT(NDAT)
      CHARACTER CITY(NDAT)*15,MON(NMON)*4,TEXT*64
      OPEN(5,FILE='TABLE2.DAT',STATUS='OLD')
      READ(5,*)
      READ(5,'(A)') TEXT
      READ(5,'(15X,12A4/)') (MON(I),I=1,12)
      DO 11 I=1,NDAT
        READ(5,'(A15,12F4.0,F6.0,F6.1)')
     *          CITY(I),(RAYS(I,J),J=1,12),AVE(I),ZLAT(I)
11    CONTINUE
      CLOSE(5)
      WRITE(*,'(1X,A)') TEXT
      WRITE(*,'(1X,15X,12A4)') (MON(I),I=1,12)
      DO 12 I=1,NDAT
        WRITE(*,'(1X,A,12I4)')
     *          CITY(I),(NINT(RAYS(I,J)),J=1,12)
12    CONTINUE
C     Replace solar flux in each column by rank order
1     WRITE(*,'(1X,A)') 'Number of month (1-12):'
      READ(*,*) MONTH
      IF(MONTH.EQ.0)STOP
      MONTH=MOD(MONTH+5,12)+1
      DO 13 I=1,NDAT
        DATA(I)=RAYS(I,MONTH)
        ORDER(I)=I
13    CONTINUE
      CALL SORT2(NDAT,DATA,ORDER)
      CALL CRANK(NDAT,DATA,S)
      WRITE(*,*) 'Month of',MON(MONTH)
      WRITE(*,'(1X,A,T27,A,T36,A,T49,A)') 'City','Rank',
     *          'Solar Flux','Latitude'
      DO 14 I=1,NDAT
        NN=NINT(ORDER(I))
        WRITE(*,'(1X,A,T19,3F12.1)') CITY(NN),DATA(I),
     *          RAYS(NN,MONTH),ZLAT(NN)
14    CONTINUE
      GOTO 1
      END
```

KENDL1 and KENDL2 test for monotonic correlations of ordinal data. They differ in that KENDL1 compares two data sets of the same rank, while KENDL2 operates on a contingency table. Sample program D13R18, for example, uses KENDL1 to look for pair correlations in our five random number routines. That is to say, it tests for randomness by seeing if two consecutive numbers from the generator have a monotonic correlation. It uses the random number generators RAN0, ..., RAN4, one at a time, to generate 200 pairs of random numbers each. Then KENDL1 tests for correlation of the pairs, and a chart is made showing Kendall's τ, the standard deviation from the null hypotheses, and the probability. For a better test of the generators, you may wish to increase the number of pairs NDAT. It would also be a good idea to see how your result depends on the value of the seed IDUM.

```
      PROGRAM D13R18
C     Driver for routine KENDL1
C     Look for correlations in RAN0 ... RAN4
      PARAMETER(NDAT=200)
      DIMENSION DATA1(NDAT),DATA2(NDAT)
      CHARACTER TEXT(5)*4
      DATA TEXT/'RAN0','RAN1','RAN2','RAN3','RAN4'/
      WRITE(*,'(/1X,A/)') 'Pair correlations of RAN0 ... RAN4'
      WRITE(*,'(2X,A,T16,A,T34,A,T50,A,/)')
     *        'Program','Kendall Tau','Std. Dev.','Probability'
      DO 12 I=1,5
        IDUM=-1357
        DO 11 J=1,NDAT
          IF (I.EQ.1) THEN
            DATA1(J)=RAN0(IDUM)
            DATA2(J)=RAN0(IDUM)
          ELSE IF (I.EQ.2) THEN
            DATA1(J)=RAN1(IDUM)
            DATA2(J)=RAN1(IDUM)
          ELSE IF (I.EQ.3) THEN
            DATA1(J)=RAN2(IDUM)
            DATA2(J)=RAN2(IDUM)
          ELSE IF (I.EQ.4) THEN
            DATA1(J)=RAN3(IDUM)
            DATA2(J)=RAN3(IDUM)
          ELSE IF (I.EQ.5) THEN
            DATA1(J)=RAN4(IDUM)
            DATA2(J)=RAN4(IDUM)
          ENDIF
11      CONTINUE
        CALL KENDL1(DATA1,DATA2,NDAT,TAU,Z,PROB)
        WRITE(*,'(1X,T4,A,3F17.6)') TEXT(I),TAU,Z,PROB
12    CONTINUE
      END
```

Sample program D13R19, for subroutine KENDL2, prepares a contingency table based on the routines IRBIT1 and IRBIT2. You may recall that these routines generate random binary sequences. The program checks the sequences by breaking them into groups of three bits. Each group is treated as a three-bit binary number. Two consecutive groups then act as indices into an 8×8 contingency table that records how many times each possible sequence of six bits (two groups) occurs. For each random bit generator, NDAT=1000 samples are taken. Then the contingency table TAB(K,L) is analyzed by KENDL2 to find Kendall's τ, the standard deviation, and the probability. Notice that Kendall's τ can only be applied when both variables are ordinal (here, the numbers 0 to 7), and that the test is specifically for monotonic correlations. In this case we are actually testing whether the larger 3-bit binary numbers tend to be followed by others of their own kind. Within the program, we have expressed this roughly as a test of whether ones or zeros tend to come in groups more than they should.

```
      PROGRAM D13R19
C     Driver for routine KENDL2
C     Look for 'ones-after-zeros' in IRBIT1 and IRBIT2 sequences
      PARAMETER(NDAT=1000,IP=8,JP=8)
      DIMENSION TAB(IP,JP)
```

```
      CHARACTER TEXT(8)*3
      DATA TEXT/'000','001','010','011','100','101','110','111'/
      WRITE(*,*) 'Are ones followed by zeros and vice-versa?'
      I=IP
      J=JP
      DO 17 IFUNC=1,2
        ISEED=2468
        WRITE(*,'(/1X,A,I1/)') 'Test of IRBIT',IFUNC
        DO 12 K=1,I
          DO 11 L=1,J
            TAB(K,L)=0.0
11        CONTINUE
12      CONTINUE
        DO 15 M=1,NDAT
          K=1
          DO 13 N=0,2
            IF (IFUNC.EQ.1) THEN
              K=K+IRBIT1(ISEED)*(2**N)
            ELSE
              K=K+IRBIT2(ISEED)*(2**N)
            ENDIF
13          CONTINUE
          L=1
          DO 14 N=0,2
            IF (IFUNC.EQ.1) THEN
              L=L+IRBIT1(ISEED)*(2**N)
            ELSE
              L=L+IRBIT2(ISEED)*(2**N)
            ENDIF
14        CONTINUE
          TAB(K,L)=TAB(K,L)+1.0
15      CONTINUE
        CALL KENDL2(TAB,I,J,IP,JP,TAU,Z,PROB)
        WRITE(*,'(4X,8A6/)') (TEXT(N),N=1,8)
        DO 16 N=1,8
          WRITE(*,'(1X,A,8I6)') TEXT(N),(NINT(TAB(N,M)),M=1,8)
16      CONTINUE
        WRITE(*,'(/7X,A,T24,A,T38,A)') 'Kendall Tau','Std. Dev.',
     *        'Probability'
        WRITE(*,'(1X,3F15.6/)') TAU,Z,PROB
        WRITE(*,*) 'Press RETURN to continue ...'
        READ(*,*)
17    CONTINUE
      END
```

SMOOFT is a subroutine for smoothing data. This is not a mathematically valuable procedure since it always reduces the information content of the data. However, it is a satisfactory tool for data presentation, as it may help to make evident important features of the data. Sample program D13R20 prepares an artificial data set Y(I) with a broad maximum. It then adds noise from a Gaussian deviate. Subsequently the data is plotted three times; first the original data, then following each of two consecutive applications of SMOOFT. You will notice that the second use of SMOOFT is almost entirely ineffectual, but the first makes a significant change in the presentational quality of the graph.

```
      PROGRAM D13R20
C     Driver for routine SMOOFT
      PARAMETER(MMAX=1024,N=100,HASH=0.05,SCALE=100.0,PTS=10.0)
      DIMENSION Y(MMAX)
      CHARACTER TEXT(64)*1
      IDUM=-7
      DO 11 I=1,N
        Y(I)=3.0*I/N*EXP(-3.0*I/N)
        Y(I)=Y(I)+HASH*GASDEV(IDUM)
11    CONTINUE
      DO 15 K=1,3
        NSTP=N/20
        WRITE(*,'(/3X,A,T16,A)') 'Data:','Graph:'
        DO 14 I=1,N,NSTP
          DO 12 J=1,64
            TEXT(J)=' '
12        CONTINUE
          IBAR=NINT(SCALE*Y(I))
          DO 13 J=1,64
            IF (J.LE.IBAR) TEXT(J)='*'
13        CONTINUE
          WRITE(*,'(1X,F10.6,4X,64A1)') Y(I),
     *          (TEXT(J),J=1,64)
14      CONTINUE
        WRITE(*,'(/1X,A)') ' press RETURN to smooth ...'
        READ(*,*)
        CALL SMOOFT(Y,N,PTS)
15    CONTINUE
      END
```

Appendix

File TABLE.DAT:

Accidental Deaths by Month and Type (1979)

Month:	jan	feb	mar	apr	may	jun	jul	aug	sep	oct	nov	dec
Motor Vehicle	3298	3304	4241	4291	4594	4710	4914	4942	4861	4914	4563	4892
Falls	1150	1034	1089	1126	1142	1100	1112	1099	1114	1079	999	1181
Drowning	180	190	370	530	800	1130	1320	990	580	320	250	212
Fires	874	768	630	516	385	324	277	272	271	381	533	760
Choking	299	264	258	247	273	269	251	269	271	279	297	266
Fire-arms	168	142	122	140	153	142	147	160	162	172	266	230
Poisons	298	277	346	263	253	239	268	228	240	260	252	241
Gas-poison	267	193	144	127	70	63	55	53	60	118	150	172
Other	1264	1234	1172	1220	1547	1339	1419	1453	1359	1308	1264	1246

File TABLE2.DAT:

Average solar radiation (watts/square meter) for selected cities

Month:	jul	aug	sep	oct	nov	dec	jan	feb	mar	apr	may	jun	ave	lat
Atlanta, GA	257	246	201	166	30	102	106	140	184	236	258	271	192	34.0
Barrow, AK	208	123	56	20	0	0	0	18	87	184	248	256	100	71.0
Bismark, ND	296	251	185	132	78	60	76	121	170	217	267	284	178	47.0
Boise, ID	324	275	221	152	88	60	69	113	164	235	284	309	191	43.5
Boston, MA	240	206	165	115	70	58	67	96	142	176	228	242	150	42.5
Caribou, ME	246	218	161	102	53	51	66	111	178	194	229	232	153	47.0
Cleveland, OH	267	239	182	127	68	56	60	87	151	182	253	271	162	41.5

```
Dodge City, KS  311 287 239 184 138 113 123 153 202 256 275 315   216  38.0
El Paso, TX     324 309 278 224 178 151 160 209 266 317 346 353   260  32.0
Fresno, CA      323 293 243 182 117  77  90 143 212 264 308 337   216  37.0
Greensboro, NC  263 235 197 156 118  95  97 134 171 227 257 273   185  36.0
Honolulu, HI    305 293 271 245 208 176 175 200 234 262 300 297   247  21.0
Little Rock, AR 270 250 214 167 118  91  96 127 173 220 256 272   188  35.0
Miami, FL       260 246 216 188 171 154 166 201 238 263 267 257   219  26.0
New York, NY    251 238 175 127  77  62  71 102 151 183 220 255   159  41.0
Omaha, NE       275 252 192 142  96  80  99 134 172 224 248 272   182  21.0
Rapid City, SD  288 262 208 152  99  76  90 135 193 235 259 287   190  44.0
Seattle, WA     242 209 150  84  44  29  34  60 118 174 216 228   132  47.5
Tucson, AZ      304 286 281 216 172 144 151 195 264 322 358 343   253  41.0
Washington, DC  267 190 196 145  75  64 101 124 153 182 215 247   163  39.0
```

Chapter 14: Modeling of Data

Chapter 14 of Numerical Recipes deals with the fitting of a model function to a set of data, in order to summarize the data in terms of a few model parameters. Both traditional least-squares fitting and robust fitting are considered. Fits to a straight line are carried out by routine FIT. *More general linear least-squares fits are handled by* LFIT *and* COVSRT. *(Remember that the term "linear" here refers not to a linear dependence of the fitting function on its argument, but rather to a linear dependence of the function on its fitting parameters.) In cases where* LFIT *fails, owing probably to near degeneracy of some basis functions, the answer may still be found using* SVDFIT *and* SVDVAR. *In fact, these are generally recommended in preference to* LFIT *because they never (?) fail. For nonlinear least-squares fits, the Levenberg-Marquardt method is discussed, and is implemented in* MRQMIN, *which makes use also of* COVSRT *and* MRQCOF.

Robust estimation is discussed in several forms, and illustrated by routine MEDFIT *which fits a straight line to data points based on the criterion of least absolute deviations rather than least-squared deviations.* ROFUNC *is an auxiliary function for* MEDFIT.

⋆ ⋆ ⋆ ⋆

Routine FIT fits a set of N data points (X(I),Y(I)), with standard deviations SIG(I), to the linear model $y = Ax + B$. It uses χ^2 as the criterion for goodness-of-fit. To demonstrate FIT, we generate some noisy data in sample program D14R1. For NPT values of I we take $x = 0.1\text{I}$ and $y = -2x + 1$ plus some values drawn from a Gaussian distribution to represent noise. Then we make two calls to FIT, first performing the fit without allowance for standard deviations SIG(I), and then with such allowance. Since SIG(I) has been set to the constant value SPREAD, it should not affect the resulting parameter values. The values output from this routine are:

Ignoring standard deviation:

```
A =    .936991      Uncertainty:    .099560
B = -1.979427       Uncertainty:    .017116
Chi-squared:       23.922630
Goodness-of-fit:    1.000000
```

Including standard deviation:

```
A =    .936991      Uncertainty:    .100755
B = -1.979427       Uncertainty:    .017321
Chi-squared:       95.690510
Goodness-of-fit:     .547181
```

```
      PROGRAM D14R1
C     Driver for routine FIT
      PARAMETER(NPT=100,SPREAD=0.5)
      DIMENSION X(NPT),Y(NPT),SIG(NPT)
      IDUM=-117
      DO 11 I=1,NPT
         X(I)=0.1*I
         Y(I)=-2.0*X(I)+1.0+SPREAD*GASDEV(IDUM)
         SIG(I)=SPREAD
11    CONTINUE
      DO 12 MWT=0,1
         CALL FIT(X,Y,NPT,SIG,MWT,A,B,SIGA,SIGB,CHI2,Q)
         IF (MWT.EQ.0) THEN
            WRITE(*,'(//1X,A)') 'Ignoring standard deviation'
         ELSE
            WRITE(*,'(//1X,A)') 'Including standard deviation'
         ENDIF
         WRITE(*,'(1X,T5,A,F9.6,T24,A,F9.6)') 'A = ',A,'Uncertainty: ',
     *           SIGA
         WRITE(*,'(1X,T5,A,F9.6,T24,A,F9.6)') 'B = ',B,'Uncertainty: ',
     *           SIGB
         WRITE(*,'(1X,T5,A,4X,F10.6)') 'Chi-squared: ',CHI2
         WRITE(*,'(1X,T5,A,F10.6)') 'Goodness-of-fit: ',Q
12    CONTINUE
      END
```

LFIT carries out the same sort of fit but this time does a linear least-squares fit to a more general function. In sample program D14R2 the chosen function is a linear sum of powers of x, generated by subroutine FUNCS. For convenience in checking the result we have generated data according to $y = 1 + 2x + 3x^2 + \cdots$. This series is truncated depending on the choice of NTERM, and some Gaussian noise is added to simulate realistic data. The SIG(I) are taken as constant errors. LFIT is called three times to fit the same data. The first time LISTA(I) is set to I, so that the fitted parameters should be returned in the order A(1) $\approx$ 1.0, A(2) $\approx$ 2.0, A(3) $\approx$ 3.0. Then, as a test of the LISTA feature, which determines which parameters are to be fit and in which order, the array LISTA(I) is reversed. Finally, the fit is restricted to odd-numbered parameters, while even-numbered parameters are fixed. In this case the elements of the covariance matrix associated with fixed parameters should be zero. In D14R2, we have set NTERM=3 to fit a quadratic. You may wish to try something larger.

```
      PROGRAM D14R2
C     Driver for routine LFIT
      PARAMETER(NPT=100,SPREAD=0.1,NTERM=3)
      DIMENSION X(NPT),Y(NPT),SIG(NPT),A(NTERM),COVAR(NTERM,NTERM),
     *           LISTA(NTERM)
      EXTERNAL FUNCS
      IDUM=-911
      DO 12 I=1,NPT
         X(I)=0.1*I
         Y(I)=FLOAT(NTERM)
         DO 11 J=NTERM-1,1,-1
            Y(I)=J+Y(I)*X(I)
11       CONTINUE
         Y(I)=Y(I)+SPREAD*GASDEV(IDUM)
```

```
        SIG(I)=SPREAD
12    CONTINUE
      MFIT=NTERM
      DO 13 I=1,MFIT
        LISTA(I)=I
13    CONTINUE
      CALL LFIT(X,Y,SIG,NPT,A,NTERM,LISTA,MFIT,COVAR,NTERM,CHISQ,FUNCS)
      WRITE(*,'(/1X,T4,A,T22,A)') 'Parameter','Uncertainty'
      DO 14 I=1,NTERM
        WRITE(*,'(1X,T5,A,I1,A,F8.6,F11.6)') 'A(',I,') = ',
     *          A(I),SQRT(COVAR(I,I))
14    CONTINUE
      WRITE(*,'(/3X,A,E12.6)') 'Chi-squared = ',CHISQ
      WRITE(*,'(/3X,A)') 'Full covariance matrix'
      DO 15 I=1,NTERM
        WRITE(*,'(1X,4E12.2)') (COVAR(I,J),J=1,NTERM)
15    CONTINUE
      WRITE(*,'(/1X,A)') 'press RETURN to continue...'
      READ(*,*)
C     Now test the LISTA feature
      DO 16 I=1,NTERM
        LISTA(I)=NTERM+1-I
16    CONTINUE
      CALL LFIT(X,Y,SIG,NPT,A,NTERM,LISTA,MFIT,COVAR,NTERM,CHISQ,FUNCS)
      WRITE(*,'(/1X,T4,A,T22,A)') 'Parameter','Uncertainty'
      DO 17 I=1,NTERM
        WRITE(*,'(1X,T5,A,I1,A,F8.6,F11.6)') 'A(',I,') = ',
     *          A(I),SQRT(COVAR(I,I))
17    CONTINUE
      WRITE(*,'(/3X,A,E12.6)') 'Chi-squared = ',CHISQ
      WRITE(*,'(/3X,A)') 'Full covariance matrix'
      DO 18 I=1,NTERM
        WRITE(*,'(1X,4E12.2)') (COVAR(I,J),J=1,NTERM)
18    CONTINUE
      WRITE(*,'(/1X,A)') 'press RETURN to continue...'
      READ(*,*)
C     Now check results of restricting fit parameters
      II=1
      DO 19 I=1,NTERM
        IF (MOD(I,2).EQ.1) THEN
          LISTA(II)=I
          II=II+1
        ENDIF
19    CONTINUE
      MFIT=II-1
      CALL LFIT(X,Y,SIG,NPT,A,NTERM,LISTA,MFIT,COVAR,NTERM,CHISQ,FUNCS)
      WRITE(*,'(/1X,T4,A,T22,A)') 'Parameter','Uncertainty'
      DO 21 I=1,NTERM
        WRITE(*,'(1X,T5,A,I1,A,F8.6,F11.6)') 'A(',I,') = ',A(I),
     *          SQRT(COVAR(I,I))
21    CONTINUE
      WRITE(*,'(/3X,A,E12.6)') 'Chi-squared = ',CHISQ
      WRITE(*,'(/3X,A)') 'Full covariance matrix'
      DO 22 I=1,NTERM
        WRITE(*,'(1X,4E12.2)') (COVAR(I,J),J=1,NTERM)
22    CONTINUE
      END
```

```
      SUBROUTINE FUNCS(X,AFUNC,MA)
      DIMENSION AFUNC(MA)
      AFUNC(1)=1.0
      DO 11 I=2,MA
        AFUNC(I)=X*AFUNC(I-1)
11    CONTINUE
      END
```

COVSRT is used in conjunction with LFIT (and later with the routine SVDFIT) to redistribute the covariance matrix COVAR so that it represents the true order of coefficients, rather than the order in which they were fit. In sample routine D14R3 an artificial 10×10 covariance matrix COVAR(I,J) is created, which is all zeros except for the upper left 5×5 section, for which the elements are COVAR(I,J)=I+J-1. Then three tests are performed.

1. By setting LISTA(I) $= 2$I for I $= 1, \ldots, 5$ and MFIT=5, we spread the elements so that alternate elements are zero.
2. By taking LISTA(I) $=$ MFIT $+ 1 -$ I for I $= 1, \ldots, 5$ we put the elements in reverse order, but leave them in an upper left-hand block.
3. With LISTA(I) $= 12 - 2$I for I $= 1, \ldots, 5$ we both spread and reverse the elements.

```
      PROGRAM D14R3
C     Driver for routine COVSRT
      PARAMETER(MA=10,MFIT=5)
      DIMENSION COVAR(MA,MA),LISTA(MFIT)
      DO 12 I=1,MA
        DO 11 J=1,MA
          COVAR(I,J)=0.0
          IF (I.LE.5 .AND. J.LE.5) THEN
            COVAR(I,J)=I+J-1
          ENDIF
11      CONTINUE
12    CONTINUE
      WRITE(*,'(//2X,A)') 'Original matrix'
      DO 13 I=1,MA
        WRITE(*,'(1X,10F4.1)') (COVAR(I,J),J=1,MA)
13    CONTINUE
      WRITE(*,*) ' press RETURN to continue...'
      READ(*,*)
C     Test 1 - spread by 2
      WRITE(*,'(/2X,A)') 'Test #1 - Spread by two'
      DO 14 I=1,MFIT
        LISTA(I)=2*I
14    CONTINUE
      CALL COVSRT(COVAR,MA,MA,LISTA,MFIT)
      DO 15 I=1,MA
        WRITE(*,'(1X,10F4.1)') (COVAR(I,J),J=1,MA)
15    CONTINUE
      WRITE(*,*) ' press RETURN to continue...'
      READ(*,*)
C     Test 2 - reverse
      WRITE(*,'(/2X,A)') 'Test #2 - Reverse'
      DO 17 I=1,MA
        DO 16 J=1,MA
```

```
               COVAR(I,J)=0.0
               IF (I.LE.5 .AND. J.LE.5) THEN
                 COVAR(I,J)=I+J-1
               ENDIF
16         CONTINUE
17       CONTINUE
         DO 18 I=1,MFIT
           LISTA(I)=MFIT+1-I
18       CONTINUE
         CALL COVSRT(COVAR,MA,MA,LISTA,MFIT)
         DO 19 I=1,MA
           WRITE(*,'(1X,10F4.1)') (COVAR(I,J),J=1,MA)
19       CONTINUE
         WRITE(*,*) ' press RETURN to continue...'
         READ(*,*)
C        Test 3 - spread and reverse
         WRITE(*,'(/2X,A)') 'Test #3 - Spread and reverse'
         DO 22 I=1,MA
           DO 21 J=1,MA
             COVAR(I,J)=0.0
             IF (I.LE.5 .AND. J.LE.5) THEN
               COVAR(I,J)=I+J-1
             ENDIF
21         CONTINUE
22       CONTINUE
         DO 23 I=1,MFIT
           LISTA(I)=MA+2-2*I
23       CONTINUE
         CALL COVSRT(COVAR,MA,MA,LISTA,MFIT)
         DO 24 I=1,MA
           WRITE(*,'(1X,10F4.1)') (COVAR(I,J),J=1,MA)
24       CONTINUE
         END
```

Routine SVDFIT is recommended in preference to LFIT for performing linear least-squares fits. The sample program D14R4 puts SVDFIT to work on the data generated according to

$$F(x) = 1 + 2x + 3x^2 + 4x^3 + 5x^4 + \text{Gaussian noise.}$$

This data is fit first to a five-term polynomial sum, and then to a five-term Legendre polynomial sum. In each case SIG(I), the measurement fluctuation in y, is taken to be constant. For the polynomial fit, the resulting coefficients should clearly have the values A(I) $\approx$ I. For Legendre polynomials the expected results are:

A(1) ≈ 3.0

A(2) ≈ 4.4

A(3) ≈ 4.9

A(4) ≈ 1.6

A(5) ≈ 1.1

```
      PROGRAM D14R4
C     Driver for routine SVDFIT
      EXTERNAL FPOLY,FLEG
      PARAMETER(NPT=100,SPREAD=0.02,NPOL=5)
      DIMENSION X(NPT),Y(NPT),SIG(NPT),A(NPOL),CVM(NPOL,NPOL)
      DIMENSION U(NPT,NPOL),V(NPOL,NPOL),W(NPOL)
C     Polynomial fit
      IDUM=-911
      MP=NPT
      NP=NPOL
      DO 11 I=1,NPT
        X(I)=0.02*I
        Y(I)=1.0+X(I)*(2.0+X(I)*(3.0+X(I)*(4.0+X(I)*5.0)))
        Y(I)=Y(I)*(1.0+SPREAD*GASDEV(IDUM))
        SIG(I)=Y(I)*SPREAD
11    CONTINUE
      CALL SVDFIT(X,Y,SIG,NPT,A,NPOL,U,V,W,MP,NP,CHISQ,FPOLY)
      CALL SVDVAR(V,NPOL,NP,W,CVM,NPOL)
      WRITE(*,*) 'Polynomial fit:'
      DO 12 I=1,NPOL
        WRITE(*,'(1X,F12.6,A,F10.6)') A(I),'  +-',SQRT(CVM(I,I))
12    CONTINUE
      WRITE(*,'(1X,A,F12.6/)') 'Chi-squared',CHISQ
      CALL SVDFIT(X,Y,SIG,NPT,A,NPOL,U,V,W,MP,NP,CHISQ,FLEG)
      CALL SVDVAR(V,NPOL,NP,W,CVM,NPOL)
      WRITE(*,*) 'Legendre polynomial fit'
      DO 13 I=1,NPOL
        WRITE(*,'(1X,F12.6,A,F10.6)') A(I),'  +-',SQRT(CVM(I,I))
13    CONTINUE
      WRITE(*,'(1X,A,F12.6/)') 'Chi-squared',CHISQ
      END
```

SVDVAR is used with SVDFIT to evaluate the covariance matrix CVM of a fit with MA parameters. In program D14R5, we provide input vector W and array V for this routine via two data statements, and calculate the covariance matrix CVM determined from them. We have also done the calculation by hand and recorded the correct results in array TRU for comparison.

```
      PROGRAM D14R5
C     Driver for routine SVDVAR
      PARAMETER(MP=6,MA=3,NCVM=MA)
      DIMENSION V(MP,MP),W(MP),CVM(NCVM,NCVM),TRU(MA,MA)
      DATA W/0.0,1.0,2.0,3.0,4.0,5.0/
      DATA V/1.0,2.0,3.0,4.0,5.0,6.0,1.0,2.0,3.0,4.0,5.0,6.0,
     *        1.0,2.0,3.0,4.0,5.0,6.0,1.0,2.0,3.0,4.0,5.0,6.0,
     *        1.0,2.0,3.0,4.0,5.0,6.0,1.0,2.0,3.0,4.0,5.0,6.0/
      DATA TRU/1.25,2.5,3.75,2.5,5.0,7.5,3.75,7.5,11.25/
      WRITE(*,'(/1X,A)') 'Matrix V'
      DO 11 I=1,MP
        WRITE(*,'(1X,6F12.6)') (V(I,J),J=1,MP)
11    CONTINUE
      WRITE(*,'(/1X,A)') 'Vector W'
      WRITE(*,'(1X,6F12.6)') (W(I),I=1,MP)
      CALL SVDVAR(V,MA,MP,W,CVM,NCVM)
      WRITE(*,'(/1X,A)') 'Covariance matrix from SVDVAR'
      DO 12 I=1,MA
        WRITE(*,'(1X,3F12.6)') (CVM(I,J),J=1,MA)
```

```
12      CONTINUE
        WRITE(*,'(/1X,A)') 'Expected covariance matrix'
        DO 13 I=1,MA
            WRITE(*,'(1X,3F12.6)') (TRU(I,J),J=1,MA)
13      CONTINUE
        END
```

Routines FPOLY and FLEG are used with sample program D14R4 to generate the powers of x and the Legendre polynomials, respectively. In the case of FPOLY, sample program D14R6 is used to list the powers of x generated by FPOLY so that they may be checked "by eye". For FLEG, the generated polynomials in program D14R7 are compared to values from routine PLGNDR.

```
        PROGRAM D14R6
C       Driver for FPOLY
        PARAMETER(NVAL=15,DX=0.1,NPOLY=5)
        DIMENSION AFUNC(NPOLY)
        WRITE(*,'(/1X,T29,A)') 'Powers of X'
        WRITE(*,'(/1X,T9,A,T17,A,T27,A,T37,A,T47,A,T57,A)') 'X','X**0',
     *              'X**1','X**2','X**3','X**4'
        DO 11 I=1,NVAL
            X=I*DX
            CALL FPOLY(X,AFUNC,NPOLY)
            WRITE(*,'(1X,6F10.4)') X,(AFUNC(J),J=1,NPOLY)
11      CONTINUE
        END
```

```
        PROGRAM D14R7
C       Driver for routine FLEG
        PARAMETER(NVAL=5,DX=0.2,NPOLY=5)
        DIMENSION AFUNC(NPOLY)
        WRITE(*,'(/1X,T25,A)') 'Legendre Polynomials'
        WRITE(*,'(/1X,T8,A,T18,A,T28,A,T38,A,T48,A)')
     *              'N=1','N=2','N=3','N=4','N=5'
        DO 11 I=1,NVAL
            X=I*DX
            CALL FLEG(X,AFUNC,NPOLY)
            WRITE(*,'(1X,A,F6.2)') 'X =',X
            WRITE(*,'(1X,5F10.4,A)') (AFUNC(J),J=1,NPOLY),'  routine FLEG'
            WRITE(*,'(1X,5F10.4,A/)') (PLGNDR(J-1,0,X),J=1,NPOLY),
     *              '  routine PLGNDR'
11      CONTINUE
        END
```

MRQMIN is used along with MRQCOF to perform nonlinear least-squares fits with the Levenberg-Marquardt method. The artificial data used to try it in sample program D14R8A is computed as the sum of two Gaussians plus noise:

$$
\begin{aligned}
\mathtt{Y(I)} &= \mathtt{A(1)}\exp\{-[(\mathtt{X(I)}-\mathtt{A(2)})/\mathtt{A(3)}]^2\} \\
&\quad + \mathtt{A(4)}\exp\{-[(\mathtt{X(I)}-\mathtt{A(5)})/\mathtt{A(6)}]^2\} + \text{noise}.
\end{aligned}
$$

The A(I) are set up in a DATA statement, as are the initial guesses GUES(I) for these parameters to be used in initiating the fit. Also initialized for the fit are LISTA(I)=I for I=1,...,MFIT to specify that all six of the parameters are to be fit. On the first call to MRQMIN, ALAMDA=-1 to initialize. Then a loop is entered in which MRQMIN is iterated while testing successive values of chi-squared CHISQ. When CHISQ changes by less than 0.1 on two consecutive iterations, the fit is considered complete, and MRQMIN is called one final time with ALAMDA=0.0 so that array COVAR will return the covariance matrix. Uncertainties are derived from the square roots of the diagonal elements of COVAR. Expected results for the parameters are, of course, the values used to generate the "data" in the first place.

```
      PROGRAM D14R8A
C     Driver for routine MRQMIN
      EXTERNAL FGAUSS
      PARAMETER(NPT=100,MA=6,SPREAD=0.001)
      DIMENSION X(NPT),Y(NPT),SIG(NPT),A(MA),LISTA(MA),
     *          COVAR(MA,MA),ALPHA(MA,MA),GUES(MA)
      DATA A/5.0,2.0,3.0,2.0,5.0,3.0/
      DATA GUES/4.5,2.2,2.8,2.5,4.9,2.8/
      IDUM=-911
C     First try a sum of two Gaussians
      DO 12 I=1,100
        X(I)=0.1*I
        Y(I)=0.0
        DO 11 J=1,4,3
          Y(I)=Y(I)+A(J)*EXP(-((X(I)-A(J+1))/A(J+2))**2)
11      CONTINUE
        Y(I)=Y(I)*(1.0+SPREAD*GASDEV(IDUM))
        SIG(I)=SPREAD*Y(I)
12    CONTINUE
      MFIT=MA
      DO 13 I=1,MFIT
        LISTA(I)=I
13    CONTINUE
      ALAMDA=-1
      DO 14 I=1,MA
        A(I)=GUES(I)
14    CONTINUE
      CALL MRQMIN(X,Y,SIG,NPT,A,MA,LISTA,MFIT,COVAR,ALPHA,
     *          MA,CHISQ,FGAUSS,ALAMDA)
      K=1
      ITST=0
1     WRITE(*,'(/1X,A,I2,T18,A,F10.4,T43,A,E9.2)') 'Iteration #',K,
     *     'Chi-squared:',CHISQ,'ALAMDA:',ALAMDA
      WRITE(*,'(1X,T5,A,T13,A,T21,A,T29,A,T37,A,T45,A)') 'A(1)',
     *     'A(2)','A(3)','A(4)','A(5)','A(6)'
      WRITE(*,'(1X,6F8.4)') (A(I),I=1,6)
      K=K+1
      OCHISQ=CHISQ
      CALL MRQMIN(X,Y,SIG,NPT,A,MA,LISTA,MFIT,COVAR,ALPHA,
     *          MA,CHISQ,FGAUSS,ALAMDA)
      IF (CHISQ.GT.OCHISQ) THEN
        ITST=0
      ELSE IF (ABS(OCHISQ-CHISQ).LT.0.1) THEN
        ITST=ITST+1
      ENDIF
      IF (ITST.LT.2) THEN
        GOTO 1
      ENDIF
      ALAMDA=0.0
      CALL MRQMIN(X,Y,SIG,NPT,A,MA,LISTA,MFIT,COVAR,ALPHA,
     *          MA,CHISQ,FGAUSS,ALAMDA)
```

```
      WRITE(*,*) 'Uncertainties:'
      WRITE(*,'(1X,6F8.4/)') (SQRT(COVAR(I,I)),I=1,6)
      WRITE(*,'(1X,A)') 'Expected results:'
      WRITE(*,'(1X,F7.2,5F8.2/)') 5.0,2.0,3.0,2.0,5.0,3.0
      END
```

The nonlinear least-squares fit makes use of a vector β_k (the gradient of χ^2 in parameter-space) and α_{kl} (the Hessian of χ^2 in the same space). These quantities are produced by MRQCOF, as demonstrated by sample program D14R8B. The function is a sum of two Gaussians with noise added (the same function as in D14R8A) and it is used twice. In the first call, LISTA(I)=I and MFIT=6 so all six parameters are used. In the second call, LISTA(I)=I+3 and MFIT=3 so the first three parameters are fixed and the last three, A(4) ... A(6) are fit.

```
      PROGRAM D14R8B
C     Driver for routine MRQCOF
      EXTERNAL FGAUSS
      PARAMETER(NPT=100,MA=6,SPREAD=0.1)
      DIMENSION X(NPT),Y(NPT),SIG(NPT),A(MA),LISTA(MA),
     *       COVAR(MA,MA),ALPHA(MA,MA),BETA(MA),GUES(MA)
      DATA A/5.0,2.0,3.0,2.0,5.0,3.0/
      DATA GUES/4.9,2.1,2.9,2.1,4.9,3.1/
      IDUM=-911
C     First try sum of two gaussians
      DO 12 I=1,100
        X(I)=0.1*I
        Y(I)=0.0
        DO 11 J=1,4,3
          Y(I)=Y(I)+A(J)*EXP(-((X(I)-A(J+1))/A(J+2))**2)
11      CONTINUE
        Y(I)=Y(I)*(1.0+SPREAD*GASDEV(IDUM))
        SIG(I)=SPREAD*Y(I)
12    CONTINUE
      MFIT=MA
      DO 13 I=1,MFIT
        LISTA(I)=I
13    CONTINUE
      DO 14 I=1,MA
        A(I)=GUES(I)
14    CONTINUE
      CALL MRQCOF(X,Y,SIG,NPT,A,MA,LISTA,MFIT,ALPHA,
     *      BETA,MA,CHISQ,FGAUSS)
      WRITE(*,'(/1X,A)') 'matrix alpha'
      DO 15 I=1,MA
        WRITE(*,'(1X,6F12.4)') (ALPHA(I,J),J=1,MA)
15    CONTINUE
      WRITE(*,'(1X,A)') 'vector beta'
      WRITE(*,'(1X,6F12.4)') (BETA(I),I=1,MA)
      WRITE(*,'(1X,A,F12.4/)') 'Chi-squared:',CHISQ
C     Next fix one line and improve the other
      DO 16 I=1,3
        LISTA(I)=I+3
16    CONTINUE
      MFIT=3
      DO 17 I=1,MA
        A(I)=GUES(I)
```

```
17      CONTINUE
        CALL MRQCOF(X,Y,SIG,NPT,A,MA,LISTA,MFIT,
     *          ALPHA,BETA,MA,CHISQ,FGAUSS)
        WRITE(*,'(1X,A)') 'matrix alpha'
        DO 18 I=1,MFIT
          WRITE(*,'(1X,6F12.4)') (ALPHA(I,J),J=1,MFIT)
18        CONTINUE
        WRITE(*,'(1X,A)') 'vector beta'
        WRITE(*,'(1X,6F12.4)') (BETA(I),I=1,MFIT)
        WRITE(*,'(1X,A,F12.4/)') 'Chi-squared:',CHISQ
        END
```

FGAUSS is an example of the type of subroutine that must be supplied to MRQFIT in order to fit a user-defined function, in this case the sum of Gaussians. FGAUSS calculates both the function, and its derivative with respect to each adjustable parameter in a fairly compact fashion. The sample program D14R9 calculates the same quantities in a more pedantic fashion, just to be sure we got everything right.

```
        PROGRAM D14R9
C       Driver for routine FGAUSS
        PARAMETER(NPT=3,NLIN=2,NA=3*NLIN)
        DIMENSION A(NA),DYDA(NA),DF(NA)
        DATA A/3.0,0.2,0.5,1.0,0.7,0.3/
        WRITE(*,'(/1X,T6,A,T14,A,T19,A,T27,A,T35,A,T43,A,T51,A,T59,A)')
     *          'X','Y','DYDA1','DYDA2','DYDA3','DYDA4','DYDA5','DYDA6'
        DO 11 I=1,NPT
          X=0.3*I
          CALL FGAUSS(X,A,Y,DYDA,NA)
          E1=EXP(-((X-A(2))/A(3))**2)
          E2=EXP(-((X-A(5))/A(6))**2)
          F=A(1)*E1+A(4)*E2
          DF(1)=E1
          DF(4)=E2
          DF(2)=A(1)*E1*2.0*(X-A(2))/(A(3)**2)
          DF(5)=A(4)*E2*2.0*(X-A(5))/(A(6)**2)
          DF(3)=A(1)*E1*2.0*((X-A(2))**2)/(A(3)**3)
          DF(6)=A(4)*E2*2.0*((X-A(5))**2)/(A(6)**3)
          WRITE(*,'(1X,A/,8F8.4)') 'from FGAUSS',X,Y,(DYDA(J),J=1,6)
          WRITE(*,'(1X,A/,8F8.4/)') 'independent calc.',X,F,(DF(J),J=1,6)
11      CONTINUE
        END
```

MEDFIT is a subroutine illustrating a more "robust" way of fitting. It performs a fit of data to a straight line, but instead of using the least-squares criterion for figuring the merit of a fit, it uses the least-absolute-deviation. For comparison, sample routine D14R10 fits lines to a noisy linear data set, using first the least-squares routine FIT, and then the least-absolute-deviation routine MEDFIT. You may be interested to see if you can figure out what mean value of absolute deviation you expect for data with gaussian noise of amplitude SPREAD.

```
        PROGRAM D14R10
C       Driver for routine MEDFIT
        PARAMETER(NPT=100,SPREAD=0.1)
        DIMENSION X(NPT),Y(NPT),SIG(NPT)
        IDUM=-1984
        DO 11 I=1,NPT
```

```
          X(I)=0.1*I
          Y(I)=-2.0*X(I)+1.0+SPREAD*GASDEV(IDUM)
          SIG(I)=SPREAD
11      CONTINUE
        MWT=1
        CALL FIT(X,Y,NPT,SIG,MWT,A,B,SIGA,SIGB,CHI2,Q)
        WRITE(*,'(/1X,A)') 'According to routine FIT the result is:'
        WRITE(*,'(1X,T5,A,F8.4,T20,A,F8.4)') 'A = ',A,'Uncertainty: ',
     *          SIGA
        WRITE(*,'(1X,T5,A,F8.4,T20,A,F8.4)') 'B = ',B,'Uncertainty: ',
     *          SIGB
        WRITE(*,'(1X,T5,A,F8.4,A,I4,A)') 'Chi-squared: ',CHI2,
     *          ' for ',NPT,' points'
        WRITE(*,'(1X,T5,A,F8.4)') 'Goodness-of-fit: ',Q
        WRITE(*,'(/1X,A)') 'According to routine MEDFIT the result is:'
        CALL MEDFIT(X,Y,NPT,A,B,ABDEV)
        WRITE(*,'(1X,T5,A,F8.4)') 'A = ',A
        WRITE(*,'(1X,T5,A,F8.4)') 'B = ',B
        WRITE(*,'(1X,T5,A,F8.4)') 'Absolute deviation (per data point): ',ABDEV
        WRITE(*,'(1X,T5,A,F8.4,A)') '(note: Gaussian spread is',SPREAD,')'
        END
```

ROFUNC is an auxiliary function for MEDFIT. It evaluates the quantity

$$\sum_{i=1}^{N} x_i \operatorname{sgn}(y_i - a - bx_i)$$

given arrays x_i and y_i. Data are communicated to and from ROFUNC primarily through the common block ARRAYS, but the value of the sum above is returned as the value of ROFUNC(B). ABDEV is the summed absolute deviation, and AA (listed below as A) is given the value which minimizes ABDEV. Our results for these quantities are:

B	A	ROFUNC	ABDEV
-2.10	1.51	245.40	25.38
-2.08	1.41	242.20	20.54
-2.06	1.30	242.20	15.69
-2.04	1.20	234.20	10.94
-2.02	1.09	193.20	6.61
-2.00	1.00	22.00	4.07
-1.98	.89	-199.20	5.78
-1.96	.79	-237.40	10.37
-1.94	.68	-246.40	15.25
-1.92	.58	-246.40	20.18
-1.90	.48	-248.40	25.13

```
        PROGRAM D14R11
C       Driver for routine ROFUNC
        PARAMETER(NMAX=1000,SPREAD=0.05)
        COMMON /ARRAYS/ NPT,X(NMAX),Y(NMAX),ARR(NMAX),AA,ABDEV
        IDUM=-11
        NPT=100
        DO 11 I=1,NPT
          X(I)=0.1*I
          Y(I)=-2.0*X(I)+1.0+SPREAD*GASDEV(IDUM)
11      CONTINUE
```

```
      WRITE(*,'(/1X,T10,A,T20,A,T26,A,T37,A/)') 'B','A','ROFUNC','ABDEV'
      DO 12 I=-5,5
        B=-2.0+0.02*I
        RF=ROFUNC(B)
        WRITE(*,'(1X,4F10.2)') B,AA,ROFUNC(B),ABDEV
12    CONTINUE
      END
```

Chapter 15: Ordinary Differential Equations

Chapter 15 of Numerical Recipes deals with the integration of ordinary differential equations, restricting its attention specifically to initial-value problems. Three practical methods are introduced: 1) Runge-Kutta methods (`RK4`*,* `RKDUMB`*,* `RKQC`*, and* `ODEINT`*), 2) Richardson extrapolation and the Bulirsch-Stoer method (*`BSSTEP`*,* `MMID`*,* `RZEXTR`*,* `PZEXTR`*), 3) predictor-corrector methods. In general, for applications not demanding high precision, and where convenience is paramount, the fourth-order Runge-Kutta with adaptive step-size control is recommended. For higher precision applications, the Bulirsch-Stoer method dominates. The predictor-corrector methods are covered because of their history of widespread use, but are not regarded (by us) as having an important role today. (For a possible exception to this strong statement, see Numerical Recipes.)*

⋆ ⋆ ⋆ ⋆

Routine RK4 advances the solution vector Y(N) of a set of ordinary differential equations over a single small interval H in x using the fourth-order Runge-Kutta method. The operation is shown by sample program D15R1 for an array of four variables Y(1),...,Y(4). The first-order differential equations satisfied by these variables are specified by the accompanying routine DERIVS, and are simply the equations describing the first four Bessel functions $J_0(x),\ldots,J_3(x)$. The Y's are initialized to the values of these functions at $x = 1.0$. Note that the values of DYDX are also initialized at $x = 1.0$, because RK4 uses the values of DYDX before its first call to DERIVS. The reason for this is discussed in the text. The sample program calls RK4 with H (the step-size) set to various values from 0.2 to 1.0, so that you can see how well RK4 can do even with quite sizeable steps.

```
      PROGRAM D15R1
C     Driver for routine RK4
      EXTERNAL DERIVS
      PARAMETER(N=4)
      DIMENSION Y(N),DYDX(N),YOUT(N)
      X=1.0
      Y(1)=BESSJ0(X)
      Y(2)=BESSJ1(X)
      Y(3)=BESSJ(2,X)
      Y(4)=BESSJ(3,X)
      DYDX(1)=-Y(2)
      DYDX(2)=Y(1)-Y(2)
      DYDX(3)=Y(2)-2.0*Y(3)
      DYDX(4)=Y(3)-3.0*Y(4)
      WRITE(*,'(/1X,A,T19,A,T31,A,T43,A,T55,A)')
     *        'Bessel Function:','J0','J1','J3','J4'
```

```
      DO 11 I=1,5
        H=0.2*I
        CALL RK4(Y,DYDX,N,X,H,YOUT,DERIVS)
        WRITE(*,'(/1X,A,F6.2)') 'For a step size of:',H
        WRITE(*,'(1X,A10,4F12.6)') 'RK4:',(YOUT(J),J=1,4)
        WRITE(*,'(1X,A10,4F12.6)') 'Actual:',BESSJ0(X+H),
     *          BESSJ1(X+H),BESSJ(2,X+H),BESSJ(3,X+H)
11    CONTINUE
      END
      SUBROUTINE DERIVS(X,Y,DYDX)
      DIMENSION Y(1),DYDX(1)
      DYDX(1)=-Y(2)
      DYDX(2)=Y(1)-(1.0/X)*Y(2)
      DYDX(3)=Y(2)-(2.0/X)*Y(3)
      DYDX(4)=Y(3)-(3.0/X)*Y(4)
      RETURN
      END
```

RKDUMB is an extension of RK4 which allows you to integrate over larger intervals. It is "dumb" in the sense that it has no adaptive step-size determination, and no code to estimate errors. Sample program D15R2 works with the same functions and derivatives as the previous program, but integrates from X1=1.0 to X2=20.0, breaking the interval into NSTEP=150 equal steps. The variables VSTART(1),...,VSTART(4) which become the starting values of the Y's, are initialized as before, but their derivatives this time are not initialized; RKDUMB takes care of that. This time only the results for the fourth variable $J_3(x)$ are listed, and only every tenth value is given. The values are passed in the named common block PATH.

```
      PROGRAM D15R2
C     Driver for routine RKDUMB
      PARAMETER(NVAR=4,NSTEP=150)
      DIMENSION VSTART(NVAR)
      COMMON /PATH/X(200),Y(10,200)
      EXTERNAL DERIVS
      X1=1.0
      VSTART(1)=BESSJ0(X1)
      VSTART(2)=BESSJ1(X1)
      VSTART(3)=BESSJ(2,X1)
      VSTART(4)=BESSJ(3,X1)
      X2=20.0
      CALL RKDUMB(VSTART,NVAR,X1,X2,NSTEP,DERIVS)
      WRITE(*,'(/1X,T9,A,T17,A,T31,A/)') 'X','Integrated','BESSJ3'
      DO 11 I=1,(NSTEP/10)
        J=10*I
        WRITE(*,'(1X,F10.4,2X,2F12.6)') X(J),Y(4,J),BESSJ(3,X(J))
11    CONTINUE
      END
      SUBROUTINE DERIVS(X,Y,DYDX)
      DIMENSION Y(*),DYDX(*)
      DYDX(1)=-Y(2)
      DYDX(2)=Y(1)-(1.0/X)*Y(2)
      DYDX(3)=Y(2)-(2.0/X)*Y(3)
      DYDX(4)=Y(3)-(3.0/X)*Y(4)
      RETURN
      END
```

RKQC performs a single step of fifth-order Runge-Kutta integration, this time with monitoring of local truncation error and corresponding step-size adjustment. Its sample program D15R3 is similar to that for routine RK4, using four Bessel functions as the example, and starting the integration at $x = 1.0$. However, on each pass a value is set for EPS, the desired accuracy, and the trial value HTRY for the interval size is set to 0.1. For the first few passes, EPS is not too demanding and HTRY may be perfectly adequate. As EPS becomes smaller, the routine will be forced to diminish H and return smaller values of HDID and HNEXT. Our results (in single precision) are:

eps	htry	hdid	hnext
.3679E+00	.10	.100000	.400000
.1353E+00	.10	.100000	.354954
.4979E-01	.10	.100000	.287823
.1832E-01	.10	.100000	.233879
.6738E-02	.10	.100000	.190423
.2479E-02	.10	.100000	.155293
.9119E-03	.10	.100000	.126883
.3355E-03	.10	.100000	.103845
.1234E-03	.10	.073460	.066323
.4540E-04	.10	.034162	.031216
.1670E-04	.10	.028686	.025915
.6144E-05	.10	.011732	.010733
.2260E-05	.10	.010758	.009784
.8315E-06	.10	.004284	.003911
.3059E-06	.10	.004460	.004035

```
      PROGRAM D15R3
C     Driver for routine RKQC
      EXTERNAL DERIVS
      PARAMETER(N=4)
      DIMENSION Y(N),DYDX(N),YSCAL(N)
      X=1.0
      Y(1)=BESSJ0(X)
      Y(2)=BESSJ1(X)
      Y(3)=BESSJ(2,X)
      Y(4)=BESSJ(3,X)
      DYDX(1)=-Y(2)
      DYDX(2)=Y(1)-Y(2)
      DYDX(3)=Y(2)-2.0*Y(3)
      DYDX(4)=Y(3)-3.0*Y(4)
      DO 11 I=1,N
        YSCAL(I)=1.0
11    CONTINUE
      HTRY=0.1
      WRITE(*,'(/1X,T8,A,T19,A,T31,A,T43,A)')
     *       'eps','htry','hdid','hnext'
      DO 12 I=1,15
        EPS=EXP(-FLOAT(I))
        CALL RKQC(Y,DYDX,N,X,HTRY,EPS,YSCAL,HDID,HNEXT,DERIVS)
        WRITE(*,'(2X,E12.4,F8.2,2X,2F12.6)') EPS,HTRY,HDID,HNEXT
12    CONTINUE
      END
      SUBROUTINE DERIVS(X,Y,DYDX)
      DIMENSION Y(1),DYDX(1)
      DYDX(1)=-Y(2)
```

```
      DYDX(2)=Y(1)-(1.0/X)*Y(2)
      DYDX(3)=Y(2)-(2.0/X)*Y(3)
      DYDX(4)=Y(3)-(3.0/X)*Y(4)
      RETURN
      END
```

The full driver routine for RKQC, which provides Runge-Kutta integration over large intervals with adaptive step-size control, is ODEINT. It plays the same role for RKQC that RKDUMB plays for RK4, and like RKDUMB it stores intermediate results in a common block named PATH. Integration is performed on four Bessel functions from X1=1.0 to X2=10.0, with an accuracy EPS=1.0E-4. Independent of the values of step-size actually used by ODEINT, intermediate values will be recorded only at intervals greater than DXSAV. The sample program returns values of $J_3(x)$ for checking against actual values produced by BESSJ. It also records how many steps were successful, and how many were "bad". Bad steps are redone, and indicate no extra loss in accuracy. At the same time, they do represent a loss in efficiency, so that an excessive number of bad steps should initiate an investigation.

```
      PROGRAM D15R4
C     Driver for ODEINT
      EXTERNAL DERIVS,RKQC
      COMMON /PATH/KMAX,KOUNT,DXSAV,X(200),Y(10,200)
      PARAMETER(NVAR=4)
      DIMENSION YSTART(NVAR)
      X1=1.0
      X2=10.0
      YSTART(1)=BESSJ0(X1)
      YSTART(2)=BESSJ1(X1)
      YSTART(3)=BESSJ(2,X1)
      YSTART(4)=BESSJ(3,X1)
      EPS=1.0E-4
      H1=0.1
      HMIN=0.0
      KMAX=100
      DXSAV=(X2-X1)/20.0
      CALL ODEINT(YSTART,NVAR,X1,X2,EPS,H1,HMIN,NOK,NBAD,DERIVS,RKQC)
      WRITE(*,'(/1X,A,T30,I3)') 'Successful steps:',NOK
      WRITE(*,'(1X,A,T30,I3)') 'Bad steps:',NBAD
      WRITE(*,'(1X,A,T30,I3)') 'Stored intermediate values:',KOUNT
      WRITE(*,'(/1X,T9,A,T20,A,T33,A)') 'X','Integral','BESSJ(3,X)'
      DO 11 I=1,KOUNT
        WRITE(*,'(1X,F10.4,2X,2F14.6)') X(I),Y(4,I),BESSJ(3,X(I))
11    CONTINUE
      END
      SUBROUTINE DERIVS(X,Y,DYDX)
      DIMENSION Y(1),DYDX(1)
      DYDX(1)=-Y(2)
      DYDX(2)=Y(1)-(1.0/X)*Y(2)
      DYDX(3)=Y(2)-(2.0/X)*Y(3)
      DYDX(4)=Y(3)-(3.0/X)*Y(4)
      RETURN
      END
```

The modified midpoint routine MMID is presented in *Numerical Recipes* primarily as a component of the more powerful Bulirsch-Stoer routine. It integrates variables over an interval HTOT through a sequence of much smaller steps. Sample routine D15R5 takes the number of subintervals I to be $5, 10, 15, \ldots, 50$ so that we can witness any improvements in accuracy that may occur. The values of the four Bessel functions are compared with the results of the integrations.

```
      PROGRAM D15R5
C     Driver for routine MMID
      EXTERNAL DERIVS
      PARAMETER(NVAR=4,X1=1.0,HTOT=0.5)
      DIMENSION Y(NVAR),YOUT(NVAR),DYDX(NVAR)
      Y(1)=BESSJ0(X1)
      Y(2)=BESSJ1(X1)
      Y(3)=BESSJ(2,X1)
      Y(4)=BESSJ(3,X1)
      DYDX(1)=-Y(2)
      DYDX(2)=Y(1)-Y(2)
      DYDX(3)=Y(2)-2.0*Y(3)
      DYDX(4)=Y(3)-3.0*Y(4)
      XF=X1+HTOT
      B1=BESSJ0(XF)
      B2=BESSJ1(XF)
      B3=BESSJ(2,XF)
      B4=BESSJ(3,XF)
      WRITE(*,'(1X,A/)') 'First four Bessel functions'
      DO 11 I=5,50,5
        CALL MMID(Y,DYDX,NVAR,X1,HTOT,I,YOUT,DERIVS)
        WRITE(*,'(1X,A,F6.4,A,F6.4,A,I2,A)') 'X = ',X1,
     *          ' to ',X1+HTOT,' in ',I,' steps'
        WRITE(*,'(1X,T5,A,T20,A)') 'Integration','BESSJ'
        WRITE(*,'(1X,2F12.6)') YOUT(1),B1
        WRITE(*,'(1X,2F12.6)') YOUT(2),B2
        WRITE(*,'(1X,2F12.6)') YOUT(3),B3
        WRITE(*,'(1X,2F12.6)') YOUT(4),B4
        WRITE(*,'(/1X,A)') 'press RETURN to continue...'
        READ(*,*)
11    CONTINUE
      END
      SUBROUTINE DERIVS(X,Y,DYDX)
      DIMENSION Y(1),DYDX(1)
      DYDX(1)=-Y(2)
      DYDX(2)=Y(1)-(1.0/X)*Y(2)
      DYDX(3)=Y(2)-(2.0/X)*Y(3)
      DYDX(4)=Y(3)-(3.0/X)*Y(4)
      END
```

The Bulirsch-Stoer method, illustrated by routine BSSTEP, is the integrator of choice for higher accuracy calculations on smooth functions. An interval H is broken into finer and finer steps, and the results of integration are extrapolated to zero step-size. The extrapolation is via a rational function with RZEXTR. BSSTEP monitors local truncation error and adjusts the step-size appropriately, to keep errors below EPS. From an external point of view, BSSTEP operates exactly as does RKQC: it has the same arguments and in the same order. Consequently it can be used in place of RKQC in routine ODEINT, allowing more efficient integration over large regions of x. For this reason, the sample program D15R6 is of the same form used to demonstrate RKQC.

```
      PROGRAM D15R6
C     Driver for routine BSSTEP
      EXTERNAL DERIVS
      PARAMETER(N=4)
      DIMENSION Y(N),DYDX(N),YSCAL(N)
      X=1.0
      Y(1)=BESSJ0(X)
      Y(2)=BESSJ1(X)
      Y(3)=BESSJ(2,X)
      Y(4)=BESSJ(3,X)
      DYDX(1)=-Y(2)
      DYDX(2)=Y(1)-Y(2)
      DYDX(3)=Y(2)-2.0*Y(3)
      DYDX(4)=Y(3)-3.0*Y(4)
      DO 11 I=1,N
        YSCAL(I)=1.0
11    CONTINUE
      HTRY=1.0
      WRITE(*,'(/1X,T8,A,T19,A,T31,A,T43,A/)')
     *        'eps','htry','hdid','hnext'
      DO 12 I=1,15
        EPS=EXP(-FLOAT(I))
        CALL BSSTEP(Y,DYDX,N,X,HTRY,EPS,YSCAL,HDID,HNEXT,DERIVS)
        WRITE(*,'(2X,E12.4,F8.2,2X,2F12.6)') EPS,HTRY,HDID,HNEXT
12    CONTINUE
      END
      SUBROUTINE DERIVS(X,Y,DYDX)
      DIMENSION Y(1),DYDX(1)
      DYDX(1)=-Y(2)
      DYDX(2)=Y(1)-(1.0/X)*Y(2)
      DYDX(3)=Y(2)-(2.0/X)*Y(3)
      DYDX(4)=Y(3)-(3.0/X)*Y(4)
      RETURN
      END
```

RZEXTR performs a diagonal rational function extrapolation for BSSTEP. It takes a sequence of interval lengths and corresponding integrated values, and extrapolates to the value the integral would have if the interval length were zero. Sample routine D15R8 works with a known function

$$F_n = \frac{1 - x + x^3}{(x+1)^n} \qquad n = 1,..,4$$

We extrapolate the vector YEST $= (F_1, F_2, F_3, F_4)$ given a sequence of ten values (only the last NUSE=5 of which are used). The ten values are labelled IEST=1,...,10 and are evaluated at XEST=1.0/IEST. A call to RZEXTR produces extrapolated values YZ, and estimated errors DY, and compares to the true values $(1.0, 1.0, 1.0, 1.0)$ at XEST=0.0.

```
      PROGRAM D15R8
C     Driver for routine RZEXTR
C     Feed values from a rational function
C     Fn(x)=(1-x+x**3)/(x+1)**n
      PARAMETER(NV=4,NUSE=5)
      DIMENSION YEST(NV),YZ(NV),DY(NV)
      DO 12 I=1,10
```

```
            IEST=I
            XEST=1.0/FLOAT(I)
            DUM=1.0-XEST+XEST**3
            DO 11 J=1,NV
                DUM=DUM/(XEST+1.0)
                YEST(J)=DUM
11          CONTINUE
            CALL RZEXTR(IEST,XEST,YEST,YZ,DY,NV,NUSE)
            WRITE(*,'(/1X,A,I2,A,F8.4)') 'IEST = ',I,'   XEST =',XEST
            WRITE(*,'(1X,A,4F12.6)') 'Extrap. Function: ',(YZ(J),J=1,NV)
            WRITE(*,'(1X,A,4F12.6)') 'Estimated Error:   ',(DY(J),J=1,NV)
12      CONTINUE
        WRITE(*,'(/1X,A,4F12.6)') 'Actual Values:     ',1.0,1.0,1.0,1.0
        END
```

PZEXTR is a less powerful standby for RZEXTR, to be used primarily when some problem crops up with the extrapolation. It performs a polynomial, rather than a rational function, extrapolation. The sample program D15R9 is identical to that for RZEXTR.

```
        PROGRAM D15R9
C       Driver for routine PZEXTR
C       Feed values from a rational function
C       Fn(x)=(1-x+x**3)/(x+1)**n
        PARAMETER(NV=4,NUSE=5)
        DIMENSION YEST(NV),YZ(NV),DY(NV)
        DO 12 I=1,10
            IEST=I
            XEST=1.0/FLOAT(I)
            DUM=1.0-XEST+XEST*XEST*XEST
            DO 11 J=1,NV
                DUM=DUM/(XEST+1.0)
                YEST(J)=DUM
11          CONTINUE
            CALL PZEXTR(IEST,XEST,YEST,YZ,DY,NV,NUSE)
            WRITE(*,'(/1X,A,I2)') 'I = ',I
            WRITE(*,'(1X,A,4F12.6)') 'Extrap. function:',(YZ(J),J=1,NV)
            WRITE(*,'(1X,A,4F12.6)') 'Estimated error: ',(DY(J),J=1,NV)
12      CONTINUE
        WRITE(*,'(/1X,A,4F12.6)') 'Actual values:    ',1.0,1.0,1.0,1.0
        END
```

Chapter 16: Two-Point Boundary Value Problems

Two-point boundary value problems, and their iterative solution, is the substance of Chapter 16 of Numerical Recipes. The first step is to cast the problem as a set of N coupled first-order ordinary differential equations, satisfying n_1 conditions at one boundary point, and $n_2 = N - n_1$ conditions at the other boundary point. We apply two general methods to the solutions. First are the shooting methods, typified by subroutines SHOOT *and* SHOOTF, *which enforce the n_1 conditions at one boundary and set n_2 conditions freely. Then they integrate across the interval to find discrepancies with the n_2 conditions at the other end. The Newton-Raphson method is used to reduce these discrepancies by adjusting the variable parameters.*

The other approach is the relaxation method in which the differential equations are replaced by finite difference equations on a grid that covers the range of interest. Routine SOLVDE *demonstrates this method, and is demonstrated "in action" by program* SFROID, *which uses it to compute eigenvalues of spheroidal harmonics. Since the program* SFROID *in Numerical Recipes is already self-contained, we need concern ourselves here only with shooting routines. For the purpose of comparison, we apply these routines to the same problem attacked with* SFROID.

⋆ ⋆ ⋆ ⋆

Subroutine SHOOT works as described above. Demonstration program D16R1 uses it to find eigenvalues of both prolate and oblate spheroidal harmonics. The oblate and prolate cases are handled simultaneously, although they actually involve two independent sets of three coupled first-order differential equations, one set with c^2 positive and the other with c^2 negative. The complete set of differential equations is

$$\begin{aligned}
\frac{dy_1}{dx} &= y_2 \\
\frac{dy_2}{dx} &= \frac{2x(m+1)y_2 - (y_3 - c^2x^2)y_1}{(1-x^2)} \\
\frac{dy_3}{dx} &= 0 \\
\frac{dy_4}{dx} &= y_5 \\
\frac{dy_5}{dx} &= \frac{2x(m+1)y_5 - (y_6 + c^2x^2)y_4}{(1-x^2)} \\
\frac{dy_6}{dx} &= 0
\end{aligned}$$

These are specified in subroutine DERIVS which is called, in turn, by ODEINT in SHOOT. The first three equations correspond to prolate harmonics and the second three to oblate harmonics. Comparing either set of three to equation (16.4.4) in *Numerical Recipes*, you may quickly verify that y_1 and y_4 correspond to the two spheroidal harmonic solutions, y_3 and y_6 correspond to the sought-after eigenvalues (whose derivative with respect to x is of course 0), and y_2 and y_5 are intermediate variables created to change the second-order equations to coupled first-order equations.

Two other subroutines are used by SHOOT. Subroutine LOAD sets the values of all the variables $y_1, \ldots, y_6$ at the first boundary, and SCORE calculates a discrepancy vector F (which will be zero when a successful solution has been reached) at the second boundary. Each of these subroutines has some interesting aspects. In LOAD, y_3 and y_6 are initialized to V(1) and V(2), values calculated in the sample program to give rough estimates of the size of the proper result. We arrived at these estimates just by looking through some tables of values. Also notice that, for example, y_1 is set to FACTR$+ y_2$DX. This is the same as saying that $y_1 =$ FACTR$+ (dy_1/dx)\Delta x$. The quantity FACTR comes from equation (16.4.20) in *Numerical Recipes*, and the term with DX comes from the fact that we placed the lower boundary X1 at $-1.0+$ DX (where DX=1.0E-4) rather than at -1.0. This is because dy_2/dx and dy_5/dx cannot be evaluated exactly at $x = -1.0$. The subroutine SCORE follows from equation (16.4.18) in *Numerical Recipes*. For example, if $N - M$ is odd, $y_1 = 0$ at $x = 0$, but if $N - M$ is even, then $y_2 = dy_1/dx = 0$.

That more or less explains things. Now, given M, N, and c^2, sample program D16R1 sets up estimates V(1) and V(2) and iterates the routine SHOOT until changes in the V are less than some preset fraction EPS of their size. Some values of the eigenvalues of the spheroidal harmonics are given in section 16.4 of *Numerical Recipes* if you want to check the results.

```
      PROGRAM D16R1
C     Driver for routine SHOOT
C     Solves for eigenvalues of Spheroidal Harmonics. Both
C     Prolate and Oblate case are handled simultaneously, leading
C     to six first-order equations. Unknown to SHOOT, these are
C     actually two independent sets of three coupled equations,
C     one set with c^2 positive and the other with c^2 negative.
      PARAMETER(NVAR=6,N2=2,DELTA=1.0E-3,EPS=1.0E-6,DX=1.0E-4)
      DIMENSION V(2),DELV(2),F(2),DV(2)
      COMMON C2,M,N,FACTR
1     WRITE(*,*) 'Input M,N,C-Squared (999 to end)'
      READ(*,*) M,N,C2
      IF (C2.EQ.999.) STOP
      IF ((N.LT.M).OR.(M.LT.0).OR.(N.LT.0)) GOTO 1
      FACTR=1.0
      IF (M.NE.0) THEN
        Q1=N
        DO 11 I=1,M
          FACTR=-0.5*FACTR*(N+I)*(Q1/I)
          Q1=Q1-1.0
11      CONTINUE
      ENDIF
      V(1)=N*(N+1)-M*(M+1)+C2/2.0
      V(2)=N*(N+1)-M*(M+1)-C2/2.0
```

```
      DELV(1)=DELTA*V(1)
      DELV(2)=DELV(1)
      H1=0.1
      HMIN=0.0
      X1=-1.0+DX
      X2=0.0
      WRITE(*,'(1X,T12,A,T36,A)') 'Prolate','Oblate'
      WRITE(*,'(1X,T6,A,T17,A,T30,A,T41,A)')
     *        'Mu(M,N)','Error Est.','Mu(M,N)','Error Est.'
2     CALL SHOOT(NVAR,V,DELV,N2,X1,X2,EPS,H1,HMIN,F,DV)
      WRITE(*,'(1X,4F12.6)') V(1),DV(1),V(2),DV(2)
      IF ((ABS(DV(1)).GT.ABS(EPS*V(1))).OR.
     *        ((DV(2)).GT.ABS(EPS*V(2)))) GOTO 2
      END
      SUBROUTINE LOAD(X1,V,Y)
      COMMON C2,M,N,FACTR
      DIMENSION V(2),Y(6)
      Y(3)=V(1)
      Y(2)=-(Y(3)-C2)*FACTR/2.0/(M+1.0)
      Y(1)=FACTR+Y(2)*DX
      Y(6)=V(2)
      Y(5)=-(Y(6)+C2)*FACTR/2.0/(M+1.0)
      Y(4)=FACTR+Y(5)*DX
      RETURN
      END
      SUBROUTINE SCORE(X2,Y,F)
      COMMON C2,M,N
      DIMENSION Y(6),F(2)
      IF (MOD((N-M),2).EQ.0) THEN
          F(1)=Y(2)
          F(2)=Y(5)
      ELSE
          F(1)=Y(1)
          F(2)=Y(4)
      ENDIF
      RETURN
      END
      SUBROUTINE DERIVS(X,Y,DYDX)
      COMMON C2,M,N
      DIMENSION Y(6),DYDX(6)
      DYDX(1)=Y(2)
      DYDX(3)=0.0
      DYDX(2)=(2.0*X*(M+1.0)*Y(2)-(Y(3)-C2*X*X)*Y(1))/(1.0-X*X)
      DYDX(4)=Y(5)
      DYDX(6)=0.0
      DYDX(5)=(2.0*X*(M+1.0)*Y(5)-(Y(6)+C2*X*X)*Y(4))/(1.0-X*X)
      RETURN
      END
```

Another shooting method is shooting to a fitting point. More explicitly, we set values at two boundaries, from both of which we integrate toward an intermediate point. For the spheroidal harmonics, we take the endpoints, in sample program D16R2, to be $-1.0+$ DX and $1.0-$ DX, and the intermediate point to be $x = 0.0$. For clarity, we considered only prolate spheroids. The calculation is similar to that in the previous sample program, except for these details:

1. There are only three first-order differential equations in DERIVS because of the restriction to prolate spheroids. (Note: the oblate case requires only that we input c^2 as a negative number.)
2. There are two load routines, LOAD1 and LOAD2, which set values at the two boundaries. At the first boundary Y(3) is initialized to V1(1), which is initially set to our crude guess of the magnitude of the eigenvalue. Y(1), the spheroidal harmonic value itself, is set to FACTR$+(dy_1/dx)\Delta x$, and Y(2) is also set as before. At boundary two, Y(3) and Y(1) are given guessed values for the eigenvalue and for $y(1-\Delta x)$ respectively. We treat the guessed eigenvalue at boundary two as independent of that at boundary one, although they ought certainly to converge to the same value. To verify this point, we make the initial guess that the values differ by 1.0 (i.e. V2(2)=V1(1)+1.0).

Sample program D16R2 otherwise proceeds much as D16R1 did, however with SCORE kept at $x = 0.0$ where the solutions must match up. The subroutine SCORE has been set to a dummy operation equating F_i to y_i so that the condition of success is that the y_i all match at $x = 0$. This is discussed more fully in *Numerical Recipes.* Check the eigenvalue results against the previous routine.

```
      PROGRAM D16R2
C     Driver for routine SHOOTF
      PARAMETER(NVAR=3,N1=2,N2=1,DELTA=1.0E-3,EPS=1.0E-6,DXX=1.0E-4)
      DIMENSION V1(N2),DELV1(N2),V2(N1),DELV2(N1),
     *          DV1(N2),DV2(N1),F(NVAR)
      COMMON C2,M,N,FACTR,DX
1     WRITE(*,*) 'INPUT M,N,C-SQUARED (999 TO STOP)'
      READ(*,*) M,N,C2
      IF (C2.EQ.999.) STOP
      IF ((N.LT.M).OR.(M.LT.0)) THEN
        WRITE(*,*) 'Improper arguments'
        GOTO 1
      ENDIF
      FACTR=1.0
      IF (M.NE.0) THEN
        Q1=N
        DO 11 I=1,M
          FACTR=-0.5*FACTR*(N+I)*(Q1/I)
          Q1=Q1-1.0
11      CONTINUE
      ENDIF
      DX=DXX
      V1(1)=N*(N+1)-M*(M+1)+C2/2.0
      IF (MOD(N-M,2).EQ.0) THEN
        V2(1)=SIGN(1.0,FACTR)
      ELSE
        V2(1)=-SIGN(1.0,FACTR)
      ENDIF
      V2(2)=V1(1)+1.0
      DELV1(1)=DELTA*V1(1)
      DELV2(1)=DELTA*FACTR
      DELV2(2)=DELV1(1)
      H1=0.1
      HMIN=0.0
      X1=-1.0+DX
      X2=1.0-DX
```

```
      XF=0.0
      WRITE(*,'(/1X,T20,A,T40,A,T60,A)') 'Mu(-1)','Y(1-dx)','Mu(+1)'
2     CALL SHOOTF(NVAR,V1,V2,DELV1,DELV2,N1,N2,X1,X2,
     *      XF,EPS,H1,HMIN,F,DV1,DV2)
      WRITE(*,'(/4X,A,3F20.6)') 'V ',V1(1),V2(1),V2(2)
      WRITE(*,'(4X,A,3F20.6)') 'DV',DV1(1),DV2(1),DV2(2)
      IF (ABS(DV1(1)).GT.ABS(EPS*V1(1))) GOTO 2
      END
      SUBROUTINE LOAD1(X1,V1,Y)
      COMMON C2,M,N,FACTR,DX
      DIMENSION V1(1),Y(3)
      Y(3)=V1(1)
      Y(2)=-(Y(3)-C2)*FACTR/2.0/(M+1.0)
      Y(1)=FACTR+Y(2)*DX
      RETURN
      END
      SUBROUTINE LOAD2(X2,V2,Y)
      COMMON C2,M,N
      DIMENSION V2(2),Y(3)
      Y(3)=V2(2)
      Y(1)=V2(1)
      Y(2)=(Y(3)-C2)*Y(1)/2.0/(M+1.0)
      RETURN
      END
      SUBROUTINE SCORE(XF,Y,F)
      COMMON C2,M,N
      DIMENSION Y(3),F(3)
      DO 11 I=1,3
        F(I)=Y(I)
11    CONTINUE
      RETURN
      END
      SUBROUTINE DERIVS(X,Y,DYDX)
      COMMON C2,M,N
      DIMENSION Y(3),DYDX(3)
      DYDX(1)=Y(2)
      DYDX(3)=0.0
      DYDX(2)=(2.0*X*(M+1.0)*Y(2)-(Y(3)-C2*X*X)*Y(1))/(1.0-X*X)
      RETURN
      END
```

Chapter 17: Partial Differential Equations

Several methods for solving partial differential equations by numerical means are treated in Chapter 17 of Numerical Recipes. All are finite differencing methods, including forward time centered space differencing, the Lax method, staggered leapfrog differencing, the two-step Lax-Wendroff scheme, the Crank-Nicholson method, Fourier analysis and cyclic reduction (FACR), Jacobi's method, the Gauss-Seidel method, simultaneous over-relaxation (SOR) with and without Chebyshev acceleration, and operator splitting methods as exemplified by the alternating direction implicit (ADI) method. There are so many methods, in fact, that we have not provided each topic with a subroutine of its own. In many cases the nature of such subroutines follows naturally from the description. In other cases, you will have to consult other references. The subroutines that do appear in the chapter, `SOR` *and* `ADI`*, show two of the more useful and efficient methods for elliptic equations in application.*

⋆ ⋆ ⋆ ⋆

Subroutine `SOR` incorporates simultaneous over-relaxation with Chebyshev acceleration to solve an elliptic partial differential equation. As input it accepts six arrays of coefficients, an estimate of the spectral radius of Jacobi iteration, and a trial solution which is often just set to zero over the solution grid. In program `D17R1` the method is applied to the model problem

$$\frac{\partial^2 u}{\partial x^2} + \frac{\partial^2 u}{\partial y^2} = \rho$$

which is treated as the relaxation problem

$$\frac{\partial u}{\partial t} = \frac{\partial^2 u}{\partial x^2} + \frac{\partial^2 u}{\partial y^2} - \rho$$

Using FTCS differencing, this becomes

$$u^n_{j+1,l} + u^n_{j-1,l} + u^n_{j,l+1} + u^n_{j,l-1} - 4u^{n+1}_{j,l} = \rho_{jl}\Delta^2$$

(The notation is explained in Chapter 17 of *Numerical Recipes.*) This is a simple form of the general difference equation to which `SOR` may be applied, with

$$A_{jl} = B_{jl} = C_{jl} = D_{jl} = 1.0 \text{ and } E_{jl} = -4.0$$

for all j and l. The starting guess for u is $u_{jl} = 0.0$ for all j, l. For a source function $F_{j,l}$ we took $F_{j,l} = 0.0$ except directly in the center of the grid where

F(MIDL,MIDL) = 2.0. The value of ρ_{Jacobi}, which is called RJAC, is taken from equation (17.5.24) of *Numerical Recipes*,

$$\rho_{Jacobi} = \frac{\cos\dfrac{\pi}{J} + \left(\dfrac{\Delta x}{\Delta y}\right)^2 \cos\dfrac{\pi}{L}}{1 + \left(\dfrac{\Delta x}{\Delta y}\right)^2}$$

In this case, J=L=JMAX and $\Delta x = \Delta y$ so RJAC $= \cos(\pi/$JMAX$)$. A call to SOR leads to the solution shown below. As a test that this is indeed a solution to the finite difference equation, the program plugs the result back into that equation, calculating

$$F_{j,l} = u^n_{j+1,l} + u^n_{j-1,l} + u^n_{j,l+1} + u^n_{j,l-1} - 4u^{n+1}_{j,l}$$

The test is whether $F_{j,l}$ is almost everywhere zero, but equal to 2.0 at the very centerpoint of the grid.

```
SOR solution grid:
   .00   .00   .00   .00   .00   .00   .00   .00   .00   .00   .00
   .00  -.02  -.04  -.06  -.08  -.09  -.08  -.06  -.04  -.02   .00
   .00  -.04  -.09  -.13  -.17  -.19  -.17  -.13  -.09  -.04   .00
   .00  -.06  -.13  -.20  -.28  -.32  -.28  -.20  -.13  -.06   .00
   .00  -.08  -.17  -.28  -.41  -.55  -.41  -.28  -.17  -.08   .00
   .00  -.09  -.19  -.32  -.55 -1.05  -.55  -.32  -.19  -.09   .00
   .00  -.08  -.17  -.28  -.41  -.55  -.41  -.28  -.17  -.08   .00
   .00  -.06  -.13  -.20  -.28  -.32  -.28  -.20  -.13  -.06   .00
   .00  -.04  -.09  -.13  -.17  -.19  -.17  -.13  -.09  -.04   .00
   .00  -.02  -.04  -.06  -.08  -.09  -.08  -.06  -.04  -.02   .00
   .00   .00   .00   .00   .00   .00   .00   .00   .00   .00   .00
```

```
      PROGRAM D17R1
C     Driver for routine SOR
      PARAMETER(JMAX=11,PI=3.1415926)
      IMPLICIT REAL*8(A-H,O-Z)
      DIMENSION A(JMAX,JMAX),B(JMAX,JMAX),C(JMAX,JMAX),
     *    D(JMAX,JMAX),E(JMAX,JMAX),F(JMAX,JMAX),U(JMAX,JMAX)
      DO 12 I=1,JMAX
        DO 11 J=1,JMAX
          A(I,J)=1.0
          B(I,J)=1.0
          C(I,J)=1.0
          D(I,J)=1.0
          E(I,J)=-4.0
          F(I,J)=0.0
          U(I,J)=0.0
11      CONTINUE
12    CONTINUE
      MIDL=JMAX/2+1
      F(MIDL,MIDL)=2.0
      RJAC=COS(PI/JMAX)
      CALL SOR(A,B,C,D,E,F,U,JMAX,RJAC)
      WRITE(*,'(1X,A)') 'SOR Solution:'
      DO 13 I=1,JMAX
        WRITE(*,'(1X,11F6.2)') (U(I,J),J=1,JMAX)
13    CONTINUE
```

```
      WRITE(*,'(/1X,A)') 'Test that solution satisfies Difference Eqns:'
      DO 15 I=2,JMAX-1
        DO 14 J=2,JMAX-1
          F(I,J)=U(I+1,J)+U(I-1,J)+U(I,J+1)
     *          +U(I,J-1)-4.0*U(I,J)
14      CONTINUE
        WRITE(*,'(7X,11F6.2)') (F(I,J),J=2,JMAX-1)
15    CONTINUE
      END
```

Routine ADI uses the alternating direction implicit method for solving partial differential equations. This method can be considerably more efficient than the SOR calculation, and is preferred among relaxation methods when the shape of the grid and the boundary conditions allow its use. It is admittedly slightly more difficult to program, and sometimes does not converge, but it is the recommended "first-try" algorithm. Sample program D17R2 uses the same model problem outlined above. When it is subjected to operator splitting and put in the form of equations (17.6.22) of *Numerical Recipes*, the coefficient arrays become

$$A_{jl} = C_{jl} = D_{jl} = F_{jl} = -1.0$$

$$B_{jl} = E_{jl} = 2.0$$

Again the trial solution is set to zero everywhere, and the source term is zeroed except at the centerpoint of the grid. As given in the text (equation 17.6.20) bounds on the eigenvalues are

$$\text{ALPHA} = 2\left[1 - \cos\left(\frac{\pi}{\text{JMAX}}\right)\right]$$

$$\text{BETA} = 2\left[1 - \cos\frac{(\text{JMAX}-1)\pi}{\text{JMAX}}\right]$$

where JMAX $\times$ JMAX is the dimension of the grid. The number of iterations 2^k is minimized by choosing it to be about $\ln(4\text{JMAX}/\pi)$. As in routine SOR, the solution for u is printed out and may be compared with the copy listed before program D17R1. Also, this solution is substituted into the difference equation and should give a zero result everywhere except at the centerpoint of the grid, where its value is 2.0. Notice that ADI makes calls to TRIDAG and requires a double precision version of that routine. For this reason we have included such a version with the sample program.

```
      PROGRAM D17R2
C     Driver for routine ADI
      IMPLICIT REAL*8(A-H,O-Z)
      PARAMETER(JMAX=11,PI=3.1415926)
      DIMENSION A(JMAX,JMAX),B(JMAX,JMAX),C(JMAX,JMAX),D(JMAX,JMAX),
     *     E(JMAX,JMAX),F(JMAX,JMAX),G(JMAX,JMAX),U(JMAX,JMAX)
      DO 12 I=1,JMAX
        DO 11 J=1,JMAX
          A(I,J)=-1.0
          B(I,J)=2.0
          C(I,J)=-1.0
          D(I,J)=-1.0
          E(I,J)=2.0
```

```
                F(I,J)=-1.0
                G(I,J)=0.0
                U(I,J)=0.0
11          CONTINUE
12      CONTINUE
        MID=JMAX/2+1
        G(MID,MID)=2.0
        ALPHA=2.0*(1.0-COS(PI/JMAX))
        BETA=2.0*(1.0-COS((JMAX-1)*PI/JMAX))
        ALIM=LOG(4.0*JMAX/PI)
        K=0
1       K=K+1
        IF (2**K .LT. ALIM) GOTO 1
        EPS=1.0E-4
        CALL ADI(A,B,C,D,E,F,G,U,JMAX,K,ALPHA,BETA,EPS)
        WRITE(*, '(1X,A)') 'ADI Solution:'
        DO 13 I=1,JMAX
            WRITE(*,'(1X,11F7.2)') (U(I,J),J=1,JMAX)
13      CONTINUE
        WRITE(*,'(/1X,A)') 'Test that solution satisfies Difference Eqns:'
        DO 15 I=2,JMAX-1
            DO 14 J=2,JMAX-1
                G(I,J)=-4.0*U(I,J)+U(I+1,J)
     *                  +U(I-1,J)+U(I,J-1)+U(I,J+1)
14          CONTINUE
            WRITE(*,'(8X,9F7.2)') (G(I,J),J=2,JMAX-1)
15      CONTINUE
        END
        SUBROUTINE TRIDAG(A,B,C,R,U,N)
C       This is a double precision version for use with ADI
        IMPLICIT REAL*8(A-H,O-Z)
        PARAMETER (NMAX=100)
        DIMENSION GAM(NMAX),A(N),B(N),C(N),R(N),U(N)
        IF(B(1).EQ.0.)PAUSE
        BET=B(1)
        U(1)=R(1)/BET
        DO 11 J=2,N
            GAM(J)=C(J-1)/BET
            BET=B(J)-A(J)*GAM(J)
            IF(BET.EQ.0.)PAUSE
            U(J)=(R(J)-A(J)*U(J-1))/BET
11      CONTINUE
        DO 12 J=N-1,1,-1
            U(J)=U(J)-GAM(J+1)*U(J+1)
12      CONTINUE
        RETURN
        END
```

Index of Demonstrated Subroutines

BOOK AND DISKETTE ORDERING INSTRUCTIONS (OUTSIDE U.S.A. AND CANADA)

Also published by Cambridge University Press are *Numerical Recipes: The Art of Scientific Computing* and example books and software that accompany it.

Numerical Recipes: The Art of Scientific Computing (ISBN 0-521-30811-9, 848 pages) is the main text and reference component of the *Numerical Recipes* package. It proceeds all the way from mathematical and theoretical considerations through to actual, practical computer routines. Thus with approximately 200 printed programs in *both* the FORTRAN and Pascal programming languages this book constitutes a complete subroutine library for scientific computation.

The example books contain FORTRAN and Pascal source programs respectively. These programs exercise and demonstrate all of *Numerical Recipes* subroutines, procedures, and functions. Each program contains comments and is prefaced by a short description of what it does and of which *Numerical Recipes* routines it exercises. In cases where the demonstration programs require input data, that data is supplied. In some cases, sample output is also shown. The example books should be valuable aids to readers wishing to incorporate procedures and subroutines and to conduct simple validation tests.

NUMERICAL RECIPES EXAMPLE BOOK (FORTRAN) (*This book*) ISBN 0-521-31330-9 192 pages
Contains sample program listings written in the FORTRAN language that demonstrate the use of each *Numerical Recipes* subroutine.

NUMERICAL RECIPES EXAMPLE BOOK (PASCAL) ISBN 0-521-30956-5 256 pages
Contains sample program listings written in the Pascal language that demonstrate the use of each *Numerical Recipes* procedure.

The programs listed in *Numerical Recipes: The Art of Scientific Computing* and *Numerical Recipes Example Book* are available in several machine-readable formats and programming languages. The diskettes listed below are available from Cambridge University Press. All versions of the diskettes are 5¼ inch double-sided/double density floppy diskettes. They operate on DOS 2.0/3.0 on IBM PC, XT, AT, and IBM compatible machines. The diskettes can save hours of tedious keyboarding, leaving users free to adapt, modify, or experiment with the programs.

NUMERICAL RECIPES FORTRAN DISKETTE V1.0 ISBN 0-521-30958-1
FORTRAN subroutines as listed in *Numerical Recipes: The Art of Scientific Computing* in machine-readable form.

NUMERICAL RECIPES PASCAL DISKETTE V1.0 ISBN 0-521-30955-7
Pascal procedures as listed in *Numerical Recipes: The Art of Scientific Computing* in machine-readable form.

NUMERICAL RECIPES EXAMPLE DISKETTE (FORTRAN) ISBN 0-521-30957-3
Demontration programs in the FORTRAN language as listed in *Numerical Recipes Example Book (FORTRAN)* in machine-readable form.

NUMERICAL RECIPES EXAMPLE DISKETTE (PASCAL) ISBN 0-521-30954-9
Demonstration programs in the Pascal language as listed in *Numerical Recipes Example Book (Pascal)* in machine-readable form.

To order the example books or latest version of the diskettes, complete the information below and mail this page or a copy to Cambridge University Press. RESIDENTS OF THE U.S.A. AND CANADA PLEASE USE THE FORM ON THE FOLLOWING PAGES.

Technical questions, corrections, and requests for information on other available formats and software products should be directed to Numerical Recipes Software, P.O. Box 243, Cambridge, MA 02238, U.S.A. Only diskettes with manufacturing defects may be returned to the publisher for replacement (no cash refunds).

To: Customer Services Department, Cambridge University Press, Edinburgh Building, Shaftesbury Road, Cambridge CB2 2RU, U.K.

Please send me:

_______ 30811-9 *Numerical Recipes: The Art of Scientific Computing* £25.00 each
_______ 31330-9 *Numerical Recipes Example Book (FORTRAN)* (**This book**) £15.00 each
_______ 30956-5 *Numerical Recipes Example Book (Pascal)* £15.00 each
_______ 30958-1 NUMERICAL RECIPES FORTRAN DISKETTE V1.0 £15.00 each
_______ 30955-7 NUMERICAL RECIPES Pascal DISKETTE V1.0 £15.00 each
_______ 30957-3 NUMERICAL RECIPES EXAMPLE DISKETTE (FORTRAN) £15.00 each
_______ 30954-9 NUMERICAL RECIPES EXAMPLE DISKETTE (Pascal) £15.00 each

Name ______________________________ (Block capitals please)

Address ______________________________

____

____

**Please accept my payment by cheque or money order in pounds sterling:* I enclose (circle one) a Cheque (made payable to Cambridge University Press)/UK Postal Order/International Money Order/Bank Draft/Post Office Giro.

**Please accept my payment by credit card:* Charge my (circle one) Barclaycard/Visa/Eurocard/Access/Mastercard/Bank Americard/any other credit card bearing the Interbank symbol (please specify).

Card No. ______________________________ Expiry date: __________

Signature ______________________________ Date: __________

Address as registered by card company: ______________________________

____

____

All prices include VAT and are subject to alteration without prior notice.

Cut along dotted line

BOOK AND DISKETTE ORDERING INSTRUCTIONS (U.S.A. AND CANADA)

Also published by Cambridge University Press are *Numerical Recipes: The Art of Scientific Computing* and example books and software that accompany it.

Numerical Recipes: The Art of Scientific Computing (ISBN 0-521-30811-9, 848 pages) is the main text and reference component of the *Numerical Recipes* package. It proceeds all the way from mathematical and theoretical considerations through to actual, practical computer routines. Thus with approximately 200 printed programs in *both* the FORTRAN and Pascal programming languages this book constitutes a complete subroutine library for scientific computation.

The example books contain FORTRAN and Pascal source programs respectively. These programs exercise and demonstrate all of *Numerical Recipes* subroutines, procedures, and functions. Each program contains comments and is prefaced by a short description of what it does and of which *Numerical Recipes* routines it exercises. In cases where the demonstration programs require input data, that data is supplied. In some cases, sample output is also shown. The example books should be valuable aids to readers wishing to incorporate procedures and subroutines and to conduct simple validation tests.

NUMERICAL RECIPES EXAMPLE BOOK (FORTRAN) (*This book*) ISBN 0-521-31330-9 192 pages
Contains sample program listings written in the FORTRAN language that demonstrate the use of each *Numerical Recipes* subroutine.

NUMERICAL RECIPES EXAMPLE BOOK (PASCAL) ISBN 0-521-30956-5 256 pages
Contains sample program listings written in the Pascal language that demonstrate the use of each *Numerical Recipes* procedure.

The programs listed in *Numerical Recipes: The Art of Scientific Computing* and *Numerical Recipes Example Book* are available in several machine-readable formats and programming languages. The diskettes listed below are available from Cambridge University Press. All versions of the diskettes are 5¼ inch double-sided/double density floppy diskettes. They operate on DOS 2.0/3.0 on IBM PC, XT, AT, and IBM compatible machines. The diskettes can save hours of tedious keyboarding, leaving users free to adapt, modify, or experiment with the programs.

NUMERICAL RECIPES FORTRAN DISKETTE V1.0 ISBN 0-521-30958-1
FORTRAN subroutines as listed in *Numerical Recipes: The Art of Scientific Computing* in machine-readable form.

NUMERICAL RECIPES PASCAL DISKETTE V1.0 ISBN 0-521-30955-7
Pascal procedures as listed in *Numerical Recipes: The Art of Scientific Computing* in machine-readable form.

NUMERICAL RECIPES EXAMPLE DISKETTE (FORTRAN) ISBN 0-521-30957-3
Demonstration programs in the FORTRAN language as listed in *Numerical Recipes Example Book (FORTRAN)* in machine-readable form.

NUMERICAL RECIPES EXAMPLE DISKETTE (PASCAL) ISBN 0-521-30954-9
Demonstration programs in the Pascal language as listed in *Numerical Recipes Example Book (Pascal)* in machine-readable form.

To order the main book, example books, or latest version of the diskettes, complete the information below and mail this page or a copy to Cambridge University Press. (Alternatively, customes may call the Press at 914/235-0300 [in N.Y. and Canada] or 800/431-1580 [in rest of U.S.A.] to place an order.) Orders must be accompanied by payment in U.S. funds or the equivalent in Canadian funds. New York and California residents please add appropriate sales tax. Prices are not guaranteed. Ordinary postage for shipping orders paid by the publisher. RESIDENTS OF COUNTRIES OTHER THAN THE U.S.A. AND CANADA PLEASE USE THE FORM ON THE PRECEDING PAGES.

Technical questions, corrections, and requests for information on other available formats and software products should be directed to Numerical Recipes Software, P.O. Box 243, Cambridge, MA 02238, U.S.A. Only diskettes with manufacturing defects may be returned to the publisher for replacement (no cash refunds).

To: Cambridge University Press, Orders Department, 510 North Avenue, New Rochelle, New York 10801.

Please indicate method of payment: check _____ , Mastercard _____ , or Visa _____ .

Name____

Address ____

____

____

Card No. ____________________________ Expiration date __________

Signature ____________________________ Date __________

Please send me:

________ 30811-9 *Numerical Recipes: The Art of Scientific Computing* $39.50 each

________ 31330-9 *Numerical Recipes Example Book (FORTRAN) (*This book*)* $18.95 each

________ 30956-5 *Numerical Recipes Example Book (Pascal)* $18.95 each

________ 30958-1 NUMERICAL RECIPES FORTRAN DISKETTE V1.0 $19.95 each

________ 30955-7 NUMERICAL RECIPES Pascal DISKETTE V1.0 $19.95 each

________ 30957-3 NUMERICAL RECIPES EXAMPLE DISKETTE (FORTRAN) $19.95 each

________ 30954-9 NUMERICAL RECIPES EXAMPLE DISKETTE (Pascal) $19.95 each

________ Please indicate the total number of items ordered

________ total price

________ tax, if applicable (NY and CA residents)

________ total enclosed

Cut along dotted line